ISBN 9780981685311

This is a work of fiction. All the characters and events portrayed in this book are fictional, and any resemblance to real people or incidents is purely coincidental.

Published in the United States and distributed to the world from the U.S. and the United Kingdom

Distributed by Ingram, Amazon.com, and Cowboy Logic Press at

www.cowboylogic.net

Edited by Renee Herbert

Illustrated by Chelsea Conlin

Special Thanks to editing help from:

Kevin Allen

Annie Viall

Roger Haller

And

The rest of the CLP forum who helped polish this work.

Dedicated to everyone who has gazed into a clear night sky—and dared to dream

Into the Dark

Escape of the Nomad

By Bill Housley

Into the Dark

Escape of the Nomad

From Earth to orbit.
From orbit to Mars.
From Mars to Centauri.
And from there...
to the stars.

Into the Dark

Escape of the Nomad

Into the Dark

Escape of the Nomad

Chapter 1

"It is brutal and barbaric!"

Ambassador Klicktah of the Crenum Alliance was nearing the end of his speech. He was usually soft spoken, but today he had to talk over the rumble of anger in the assembly hall. Some of the delegates agreed with him, some didn't, all were outraged.

He wished he'd remembered to bring along some hearing protection. It was good that the ancient founders of The United Council for Peace had designed the outside of the meeting hall to resemble a living planet—complete with clouds, oceans, and continents—to remind everyone, delegates and functionaries alike, what they were there to protect. Unfortunately, at times like this, the spherical shape caused the delegates' thunderous shouting inside the hall to focus toward the center like gravity and reverberate with a crushing din.

Klicktah had expected his remarks to be provocative—the treatment of grounded worlds had always split the UCP. Its founders had designed it to trap starfaring cultures in a web of common interests, but this topic often caused the delegates to forget their common interests—at least temporarily.

"Order!" the deep, booming voice of Avin Shackree, President of the UCP, rose above the clamor. As the leader of the Council, he did his job well and usually kept these exchanges fair and civil.

"We will have order here! Please let Klicktah speak!" He pounded his gavel and the noise in the room subsided.

"There, that's better. Please continue Ambassador."

"Thank you, Mr. President. We all stood here in this hall, not long ago, and agreed that we would not tolerate such behavior again," Klicktah resumed in a more diplomatic tone, "We should resolve as a council to pass a new opinion, promising consequences…"

"It is none of the UCP's or the Crenum's' business what we do within our own borders!" Ambassador Trelleg, of the Zom Republic, interrupted. "Why can't the Crenum Alliance just keep to its own affairs? Why does this council tolerate Crenum incursions onto our planets? Why doesn't the UCP limit its own activities to member and neutral worlds? Telis 3 is outside the boundaries of the Crenum Alliance and its culture is outside the jurisdiction of the UCP."

"Trelleg, you are out of order, you will get your turn to speak," Shackree replied.

"Telis 3 is grounded," Klicktah now spoke directly at Trelleg. "It has a native, non-space-faring people inhabiting it. You know that! It puts them under the governance of the UCP, regardless of where they are."

"The Mistaac Accord only protects advancing cultures," Trelleg fired back. "The people of Telis 3 are permanently grounded! They have abandoned spaceflight, and have no plans to pursue it further. Their society is in decline. They no longer have the public will or even the capacity to become a starfaring people!" He lowered his voice to a more reasonable tone. "If they were exploring interstellar space, they would have their own membership and representation on this council; if they were progressing, they would have the protection of the Mistaac Accord. As it stands they are a vassal world of the Zom Republic and their disposition is nobody's business but ours."

Into the Dark
Escape of the Nomad

"The Zom claim that the Telesians are not progressing," Klicktah continued, addressing the council again, "but our findings show otherwise."

"Your findings," said Ambassador Trelleg with a humorless smile, "are illegal. However, the Zom Republic would be very interested in anything that your spies have learned about Telesian technology. Do you have their reports with you?"

Klicktah removed a data module from a pocket in his coat and inserted it into a slot in his computer console.

He accessed the file, broadcasting it to the other delegates' terminals and displaying it on the huge, 3-D holo-image in the center of the hall.

"The Telesians have been using a primitive form of anti-gravity in various aircraft designs—and to enhance and improve some of their other vehicles, and other devices—for the past fifteen of their years. Many of these technologies have historically been precursors to star flight propulsion on other planets. For this reason and others, our researchers believe that there is a starship under construction somewhere on the planet." Moving images of various aircraft and other vehicles and tools—seen hovering and flying without thrusters or wings—appeared on the holo-image.

"Preposterous!" yelled Trelleg. He waved his hand at the display. "Those primitives have been using anti-gravity in these same disgustingly limited ways, with very little advancement in form, for that entire fifteen years."

"Again, Mr. Trelleg," President Shackree said, without taking his eyes off of the image in front of him, "You will get your chance to refute Ambassador Klicktah's data."

Trelleg made an impatient sound.

"But it's a waste of time, our own scientists have been through all of this," he argued. "They've studied these images, tracked their progress, and haven't found much new advancement since then. The people of Telis 3 have stagnated technologically, their industries continue to use mostly hydrocarbon chemical reaction and nuclear fission for most of their energy needs, and the natural resources that support those technologies are nearly depleted. There is a lot less advancement going on there than Klicktah is implying, and the Telesian governments have outlawed publicly financed space flight research. There are no government funded deep space projects, no extra-orbital probes, no manned spacecraft, no private space programs, nothing."

"That's not true, Trelleg," Klicktah replied patiently, "There is more going on there than you're revealing." He had to struggle with his tone; Trelleg's obvious attempts to rewrite the facts annoyed him. He made his next point with emphasis, drawing a breath before continuing. "Our findings show direct evidence of an independently funded project—a prototype spacecraft under construction with an artificial-gravity based propulsion system."

The murmuring and whispers that had begun in the chambers suddenly ceased.

Now you've gone and done it; you and your big mouth may just have spoiled everything, Klicktah thought to himself. He had to try to protect Telis 3, but he had to do it without revealing information that would doom the advancements currently taking place on the planet. The Zom knew that the development of artificial gravity propulsion, by any Telesian, would stop their plans for the Telis system—so such projects had a way of disappearing. Klicktah knew that such a prototype existed. Crenum scientists had seen it, even touched the hull of the craft with their hands—

Klicktah had seen the photographs. He thought about this and chewed his lip as he opened the truth another crack.

A new image appeared; it was an official looking Telesian document. The heading read:

National Aeronautic and Space Administration

Jet Propulsion Laboratory

PAGA (Projected Artificial Gravity Attenuator)

Astronaut Stan McPherson presenting

On each of their displays, the delegates could see the same document, but in their own languages. Klicktah zoomed in on a sketch of a spherical craft on page 5 of the document. The ship had a large, horizontal, ring-shaped structure encircling its middle, prompting an audible gasp from some of the delegates. Then Klicktah's aged finger appeared, giant sized, on the holo-display. He pointed to the ring.

"This proposal was presented to a government organization called 'NASA', in a country on Telis 3 called the 'United States of America' or 'U.S.A.' for short," he said. "Look familiar? The ring structure that you see encircling this craft is recognizable to everyone here, it is a dual-node, parabolic artificial gravity projector—a drive ring. This report, together with the details of this proposal, indicates that the ship in this sketch is a spacecraft similar in design and abilities to the first interstellar vehicles that each of our cultures used, to lift us out of our grounded years. In each interstellar culture, spacecraft very similar to this one began a revolution of new technologies that started the rebirth of each of our races, and eventually brought us to join hands between the stars and form this organization, the UCP. This document and sketch were produced by native scientists on Telis 3 and the ship in this drawing is currently under construction on the planet."

Klicktah stopped speaking. He was out of time.

"Trelleg, would you like to offer a rebuttal?"

"Of course, Mr. President." He took control of the council's 3-D display and pointed to some text in the upper corner of the document, it said:

July 17, 2022

"The date on this document indicates that it was in the Telesian year 2022 that this Mr. McPherson presented his team's findings to the organization called 'NASA'. Today's date, by Telesian reckoning, is…let me see…" There was a pause while he calculated something on a separate device which he held in his hand. "March 13th, 2035. Therefore, according to this document, McPherson gave this presentation almost thirteen full cycles ago. What my esteemed colleague, Klicktah, has failed to mention—in his rather over-dramatized speech—is that this 'NASA' was disbanded sometime in 2023 and that all of the research and designs for this spacecraft were lost in a radiation accident in 2025. The publicly elected government there has since outlawed all funding for extra-orbital space flight development and has been oppressively regulating and taxing all private spaceflight enterprises. They do this, as they like to put it, 'to reimburse the public' for all the money and other resources that were 'squandered' in the past, on 'worthless space adventures'."

He pointed at the display. "This document that Klicktah has presented to us with such flair is already known to the Zom; I have a copy of it in my office. These people arrogantly rejected the technologies that this scientist pioneer, this Mr. McPherson, proposed, and canceled all funding for NASA. Not just the U.S. Government either, but this attitude is a plague, effecting every government on the planet. They have all jettisoned space science funding from their public policy, in favor of pressing issues more planet-side. Serious public interest in anything off planet has been completely lost and the vision of space sciences even became unpopular in their fiction literature. Other than a few die-hard individuals with telescopes, the only attention they pay to anything outside their atmosphere is orbiting probes that look inward at the planet. The evidence that Klicktah has presented here is worthless! The scientists that participated in this proposal have all been discredited and are now employed in other pursuits. Now, even private spaceflight enterprises quickly fail because of restrictive laws, abusive taxation, and an overall lack of interest by the Telesian people.

"Furthermore, even if someone on Telis 3 were able to build a spacecraft capable of interstellar flight, building a space flight industry would be impossible for these people. They've used up nearly all of their geologic energy reserves, and since they are no longer able to gather Helium3 fuel from the reserves on their moon, and other places in the Telesian system, they don't have enough sustainable energy sources to support their population. Even hydrogen fusion energy is becoming difficult for them to maintain, because of their reduced manufacturing capacity. The industrial period on Telis 3 is dying, all of its advanced industries are in decline, and the people there are struggling just to move about, educate, and feed themselves. The age of advancement is history for them.

"Mr. Klicktah has dramatically told us that there is a private project at work; but he has shown us no physical evidence to back up his claims. Are we to take his word for it? He represents the Crenum Alliance, a nation that has been illegally trespassing on Zom space, spying on us in Telesia and elsewhere for years. Yet even with all of the information they have gathered, Klicktah still seems unable to produce even a shred of proof, not even an image, of this alleged hyper-stellar spacecraft. Zom scientists have been studying the Telesians very closely, and if such a project exists, then how could it have gone on right under our noses, and under the noses of the Telesian authorities, without being seen? How is it that only Crenum spies are aware of it? Could it be that *they* are actually the ones who are building this craft?"

A murmur began in the hall. Trelleg smiled at this and continued.

"Klicktah wants this council to oversee Zom activities on the planet, but who is overseeing Crenum activities? This little sketch that Klicktah has shown us today is conveniently similar to our own technology, and seems to imply that Telis 3 is developing interstellar sciences. Klicktah wants to build hope that the Telesians might be one of us someday. It is a very romantic notion, but who is to say that this drawing was not prepared under the coaching of Crenum agents? How can anyone be sure?"

"Outrageous!" Klicktah was on his feet, he could not believe Trelleg's audacious…

"Please sit down, Klicktah," President Shackree admonished him.

"Our researchers have built a well documented tradition studying primitive cultures, and we have always done it without interfering with them; but while we watch grounded worlds develop into starfaring societies, the Zom *Empire* covertly undermines, invades, and destroys them! Trelleg's naked hypocrisy in making this preposterous allegation is more shocking even…than his government's brazen encroachments!"

President Shackree pounded his gavel, but Klicktah continued.

"It is absurd and laughable that anyone representing the Zom could stand there, and with a straight face, accuse anyone else of interfering with the natural development of anything." Klicktah added, calming down a little. "The fact that Trelleg would make such a ridiculous allegation articulates Zom arrogance more eloquently than I ever could."

"Mr. Klicktah is trying to promote hysteria with this outburst," Ambassador Trelleg continued. "He clearly has no current, detailed data to substantiate his claim that the Telesians are progressing toward star travel. If this antiquated document, which his spies could only have found in a museum somewhere, is the extent of his evidence, then he has no evidence at all. If there is a spacecraft under construction on Telis 3, then the Zom Republic would certainly like to see it."

"So you can crush it!"

"No! So we can examine it for foreign tampering!"

"Gentlemen, gentlemen please," President Shackree cut in again. "Ambassador Trelleg, you know as well as I do that if even one native person on that planet has recently produced any fruitful spaceflight research, or is constructing anything like the spacecraft in that sketch, that we would probably find your planned occupation and biology mining of Telis 3 to be illegal.

In addition, the fact that Telis 3 lies within your borders makes no difference and you know it. That culture is far older than the Zom Republic, far older even than this council. Most of these planetary societies will last longer than our cheap copies of them. That is the reason why it is our duty to protect the rich cultures and governments on these worlds—they are the building blocks of everything else we do. It is our public policy that they rein supreme; we are here to serve them.

"It is in everyone's best interest that we preserve them and not interfere with them, and you cannot undertake an encroachment of the size that you intend without a full scale invasion of the planet. Also, I refuse to believe that your industries will not enslave whatever will be left of that planet's population to provide the labor force for such an endeavor."

He turned to Klicktah, "However, you know that there are degenerate planets whose cultures and societies have already peaked. The

Mistaac Accord does not protect such and the Zom and others are free to colonize, or otherwise make use of any such resources within their borders. The legal language that allows for this is not popular with everyone here, but the law binds us all. Now, are you prepared to dispute Trelleg's claims regarding Telesian laws and public attitudes? I could make him produce evidence of them."

"No," Klicktah shook his head, "those facts are not in dispute at this time. What *we* dispute is Trelleg's *very* subjective projections on the planet's *future* industrial capacity..."

"...and you dispute it on even more subjective grounds since you, officially, won't even admit that you have any scientists in Telesia," Shakree interrupted, "and if we send a fact finding committee, they won't be there to explore the hypothetical. I should also think that the team would have to include Zom representatives and they would need to know where to find the Telesian spacecraft, so they can examine it."

Klicktah looked down at the age spots on his hands and sighed. He had already served longer in his office than had any other Crenum Alliance delegate in history, and it had been a good run. Though he was fit for his age, he suddenly felt very old. He wondered for a moment what he would be doing right now if he had retired last year when his wife had urged him to.

She had often said that he kept himself cooped up indoors too much. Being from a farm, it had always bothered her that her husband spent so much time breathing cycled air, either inside the UCP building with his high-stress occupation or on board a spacecraft; she said that it was unhealthy. Looking around him now, he could see her point. His apartment/office was here at the all-enclosed UCP facility, his life as the Crenum Ambassador to the UCP was here, as were the various recreational activities that he engaged in—when he had the time. His duties permitted him to return home only when the council was in recess; and then he would travel for three weeks to his home planet, where he would stay for three weeks and then return here and back to work. In the early days of his assignment, his wife used to join him on Cartlar 3 when the Council was in session, but there was really nothing for an old farm girl to do in such a place, particularly when all of their children and grandchildren lived near their home on Crenumar 4.

The Crenum worlds were running out of peaceful solutions to the Zom issue. There was a war brewing; it was going to start on his watch and he was feeling increasingly powerless to prevent it. Telis 3 was very near the borders of Crenum space and the various implications of that, whether real or imagined, were adding tensions to the issues at home. The Crenum people wanted the UCP to do something about the Zom's plan to invade and occupy Telis 3, not only because the Mistaac Accord was a UCP law, and it fell to the UCP to enforce it, but also because the Crenum Alliance was not equipped to deal with the Zom, on that level, without support. However, if the Zom invaded that planet, such an act of barbarism against a defenseless, primitive world so close to the Crenum borders, might cause the planets of the Crenum Alliance to rise up and demand war against the Zom Republic. The people of the Crenum Alliance feared the Zom, and fear is a dangerous thing.

The Crenum and the Zom had the most powerful forces in the region. Each had a long list of friends and allies and had been engaged in an arms race with each other for as long as anyone could remember, but the Zom had the edge. Distance, better technology, and issues which had drawn the Zom's attention to other regions of their space, had kept the Crenum and Zom forces separated up until now, but the Telis 3 debate was proof that those days were long over.

There was a man on Telis 3 who was building a spacecraft, one that was equipped with a primitive interstellar drive. However, his own people were hostile to his efforts and he was in constant jeopardy of losing it all, because he had to do his work outside the knowledge and laws of his government. Crenum operatives on the planet had visited the man's secret facility, while he was away, and collected some very specific data; and the evidence they had gathered was clear and definitive. If Klicktah used this evidence it would assure further debate, and possibly get the UCP to prevent further Zom activities there, but he dared not risk it. All the Zom would have to do, once they knew the space craft's whereabouts, would be to find an excuse to delay the UCP team for a little while and arrange an accident.

What's more, while the Crenum field agents claimed that the Zom were in violation of the Mistaac Accord with respect to Telis 3, those same agents were actually skirting the thin edges of it themselves.

Chapter 2

Klicktah was almost frantic. It had taken his boss, Interstellar Relations Chief Gimlar, too long to call him back—and the matter was urgent.

"Thank you for your report, Mr. Klicktah, you did well. Prepare stage two of our disclosure."

"Yes, Sir, but I'm concerned about Zom reaction to news of this project—I received word that the Zom ambassador contacted his military attaché immediately after the session. I think they will launch their own investigation and search for the craft."

"Of course they will, Jeri, wouldn't you?"

"Well, yes…but…"

"We will take steps to protect Mr. McPherson and his ship. I'll use this as an excuse to step up the timetable to first contact; it'll work out for the best, my friend, don't worry."

"Very well," he sighed.

"Do you have everything you need?"

"Yes, the package arrived yesterday."

"Good, waste no time, Ambassador. We have to pinch them in very carefully; slow them down so they can take no drastic action. We cannot stop them militarily…yet…but we must not allow them to invade that planet. It's too bad the UCP General Assembly wasn't more cooperative, we could have gotten things done quicker. I guess we expected it though, didn't we?"

"Yes."

Klicktah ended the communication. He opened a drawer and removed a diplomatic communiqué packet, unsealed it, and shook out a data module. He had heard rumors, awful things, about Zom activities in Telesia. The Crenum had been collecting information on them for several years, and little bits of disturbing news had leaked out from time to time. This communiqué contained a copy of the mission's latest report on their findings. He plugged it into his office console, and then spent some time laboring through the various security seals attached to it.

"Stage two" meant launching an official complaint with the legal office of the UCP. The UCP General Advocate, whom he would have to deal with, was not known for having a pleasant disposition. She maintained no loyalties except to the UCP, which she protected as vigorously as some animals protect their young. Many a self-serving delegate had left her office with nothing but refusals and partial hearing loss. There was a joke going around that she occasionally distributed bruises as well.

"Probably just a myth," snorted Klicktah with a smile, as he waited for the files to decrypt, "a professional like her would be much too civilized for that."

Once the documents lay open before him, he reviewed them carefully. The Telis 3 operations' commander had taken extra pains to be thorough. Everything Klicktah had heard was true, and the report spelled it all out, stark and clear. Many of the methods the Zom used were clearly illegal; many others were barely legal, and some were brutal and frightening. Klicktah shook his head as he scanned through the data, he almost couldn't believe he was reading about the behavior of a civilized people.

Whoever this scientist was who was building a spacecraft on Telis 3, Klicktah pitied him. The man was in terrible danger.

It also appeared, from the details in the report, that the Zom had far closer ties with U.S. Government than he'd suspected. That meant that the Zom might pass the news about the spacecraft to the local authorities who, unlike the Zom, would know precisely how to find it.

He wondered how Gimlar and Grandin would arrange first contact, and win Macpherson's trust.

There was no other way.

***

"Stan McPherson?"

Stan looked up from his newspaper, into the eyes of someone he had never expected to see. It gave him a start. He'd been sure that aliens from space existed, perhaps that they even visited Earth secretly, but he never expected such a guest to just walk up to him like this, by the pool, like some kind of tile salesman.

Stan looked at him carefully. The alien resembled any human, but of some race that didn't exist. If it were not for the slightly too pale face, unusually shaped eyes, and sharp features, he might actually have mistaken him for a native of Earth. It was as if he belonged to some un-discovered race, hidden through the millennia of history, until the ground chose that particular moment to open up and spew him out onto Stan's patio. If he at least had had some kind of growth sticking out of the side of his face or something, it might have seemed less weird.

It made Stan wonder if maybe he and this being had a common ancestry somewhere. Maybe Earth or this alien's home-world had an older history that tied them together in origins. It was also possible that this was an example of parallel development, the theory that successful evolution always ends up taking similar paths.

The alien's clothing didn't look strange either, and Stan doubted they came from the stars. The being was dressed simply in a black Armani suite, and carried a briefcase.

"I'm Stan McPherson," he said calmly, swallowing down the fear that had rose up from his stomach. He wanted to maintain the friendly, formal atmosphere that the being in front of him seemed to want to convey.

"As you can see, I am not from your world."

"Ye…yes", Stan swallowed again, "What do you want?" *That was a poor choice of words. Shouldn't I say something a little friendlier, instead of, "Ya? wha'd'ya want"? I may just as well have said, "Wel-come to Earth…now go home!"*

"I understand that my presence here must come as somewhat of a shock," said the alien with great politeness. "Please forgive the intrusion, but urgency required it."

"It…It is a shock", said Stan, then he realized that he was staring and composed himself.

"Sorry," he motioned for his visitor to have a seat. "I knew this would happen…someday…just not here."

"Not here," said the alien with a grin, as he pointed up at the sky, "but up there, right?"

Stan said nothing.

"Well you are quite correct of course, this is not how it is normal-ly done," he sat down sideways across the sunbathing chair adjacent to Stan's. "We had intended to wait until after your ship was completed."

The alien stopped and stared for a moment at a moth struggling in the water of the pool, causing Stan to look too. Its wings beat a rhythmic pattern of ripples in the otherwise smooth surface, which then reflected back on themselves to form a pulsing resonance bulge just behind the insect.

The effect reminded Stan of how an artificial gravity drive formed a similar gravity “bulge” to push against.

Stan and a number of scientists at NASA's Jet Propulsion Laboratory had figured out how to build this fake gravity well. They'd even proven, mathematically, that if used in concert with the even stronger gravity of a giant planet, like Jupiter or Saturn, this technology could even fling a ship beyond the speed of light.

They'd known it would work, proven it on paper and in computer simulations, even presented their findings; but the project had been scrapped before they could construct a working model.

He was still staring at the moth, now spinning out of control in the water, when it hit him; the alien had spoken of his ship. He pulled his eyes away from the pool.

When his father had died, his aging mother had given him the family estate and fortune, and with it, the chance to continue the project that NASA had abandoned.

“What ship?” he asked, trying to sound convincing. The alien had stopped watching the insect and was watching Stan instead.

“Why your ship of course,” he said, grinning. He glanced back at the moth, then back at Stan.

“I'm afraid I don’t know what you're talking about.” This creature knew about his ship! His mind struggled with this; the craft wasn’t complete yet, not even able to fly. It was vulnerable. It would be a while yet before he would be able to move it, and a very long time before it would be space-worthy. If the secret got out too soon…

Still smiling, the alien set his briefcase on his lap and turned it around to face Stan; then he opened it, so that Stan could see inside. Stan's heart jumped.

Inside was a small, black object, the size and shape of a golf ball. The sphere was wedged inside a doughnut-like ring, and someone had carefully coiled and tied the long, tail-like antenna alongside it.

Stan stared at it. In his mind's eye he could see what was inside the probe, as if he was looking at it through x-ray eyes. One of the first things he had done after leaving NASA had been to build a few of the small, working test models that his team had originally planned; tiny probes to test and tune the new PAGA drive technology. The prototypes had performed well and Stan, unable to sit still, had been launching them deeper and deeper into space. Most of these probes had returned, and the data and photos they'd brought back had fired his soul.

His own hands had built the probe the alien was now showing him. He remembered carefully arranging the computer, the power cell, the camera, and the rest to all fit together within the tiny craft. He had built his probes small like this, to prevent their detection by radar and graviton sensors. He'd built each probe by hand, and each was a little different from the next as the design evolved.

He'd started out by sending them to known places, so he could test their navigation, measure their speed, check for the effects of time dilation, etc. When he grew bored with studying things that humanity already knew, he started using the probes to feed his thirst for exploration. Through the eyes of his probes, he'd chased comets through the Oorte Cloud, landed on every one of the moons of Jupiter, and even successfully navigated between the close orbits of Pluto and Charon. He'd even intercepted and photographed each of NASA's old Voyager probes.

But, in the end, he still felt empty. Before joining JPL he'd flown as an astronaut on NASA's only Mars mission, and the bug had never left him. Intellectually, he was a scientist, but in his heart, he was still an explorer. He was a man of space, in a world without the hope of space travel. He was an albatross with a broken wing. So, to satisfy his thirst, he'd started to use Jupiter's gravity well to send his probes even further. He'd sent them to map the Proxima Centari system, and even watched the finger of God forming worlds in the dust disk around the more distant star AU Microscopii.

On the more practical side, he'd sent probes to the gas giants of Lalande 21185—in the Big Dipper. From there his probes had mapped that system. What he wanted to find was an unoccupied, habitable world to which he could move his ship and finish it once it was space-worthy.

He'd sent the probe in the briefcase to photograph and radio map the third planet of Epsilon Eridani, but it hadn't returned. He remembered holding it in his hands for the last time before tossing it into a moonless sky.

Stan stood up. He quickly closed the lid of the briefcase, looking around as if the very air around them had eyes. "Not here," he hissed. "Come inside, please, quickly."

They walked to the house, closed the door, and then Stan turned to the alien, who was still smiling.

"Alright!" he said, "So you found my probe, so what?"

"Actually, it found us. By the way, we analyzed its programming, there is nothing wrong with it; it functions perfectly. A bit primitive of course, but that is to be expected. It saw some things that it should not have, so we had to wipe its logs and images, but you may have it back." He held out the briefcase.

"And somehow you think that this means I'm building a...!" Stan shouted, then he lowered his voice, "...a ship?" He ignored the briefcase.

The alien just continued looking at him, smiling.

"Who are you and what is it you said you wanted?" Stan demanded.

"We—or rather I and the people whom I represent—know about the tragic fate of your team's work at NASA. We know what they did to your research, we know what you have been doing since, and we want to help. As I may have said before, contacting you in this way is not completely...umm...legal. Our laws even forbid us from allowing probes like this one to report back with their findings, if we can help it. About the only correct way for us to make contact with you would have been to leave you alone until you finished your spacecraft and met you in space; but that would never have happened."

He leaned close and lowered his voice.

"At the rate you are going, you will never get your ship completed. Forces you can't see are being set in motion to stop you. You are on the right track, and have had a real shot at success, and we have been waiting, patiently, but now your work is about to be lost again, this time forever."

Stan felt a lump forming in his throat. He had seen signs that certain agencies had noticed his purchases and had been making inquiries into his activities.

"Even when you are sure you can safely take the ship outside the atmosphere, it would be very hazardous for you to go snooping around out there right now. There are currently very few safe havens for you, and your solar system is being watched. Other people—people not from your world—will intercept you and end your work."

"My plan was to find an uninhabited system..."

"There are no uninhabited systems," the alien said. "There are no unexplored worlds for you to discover—and the only naked savage out there would be you. That is why I am here. We saw a need to contact you early, to help you speed the work along and try to protect you. Then, when your ship is ready, we will guide you to a place of safety."

Stan felt sick.

"Your government is able to find your ship by tracing your family's financial holdings. They have already suspected you, this you probably already know, but what you don't know is that there have been some new developments—they are going to become more...motivated...to find your project and dismantle it. Some people outside your planet have taken an interest in you, and are about to work through your government to learn your ship's location. They may even pressure your own people to move against you themselves. You and your ship are in the way of some very bad people who are right now turning over every stone looking for it. Those whom I represent want...to help ready your ship for flight, and then escort it past those who wish to destroy it. I cannot tell you more yet, only that we need your help, and we know that you need ours."

***

CIA director Bill Jones was sitting in his office, holding a large photograph, when a knock came to his door.

"Come!"

Agent Rod Mills entered the room; the director pointed to a chair next to his desk.

"Sit," he said.

Agent Mills sat down and Director Jones slid the photo across his desk to him. Mills looked at it and sighed.

"You know what this means?" Jones said.

Agent Mills looked up.

"Yes I do."

"The question is what do we do now?" Director Jones shook his head. "They sent a guy to his house Rod, and from the way he's dressed, it doesn't look like he went there for a swim."

Agent Mills looked up from the photo, "This may not be one of our aliens at all; maybe it's someone else entirely."

The CIA director pondered that for a moment. "They said that Earth is in an area of space that is controlled by them and them alone, and that other aliens were forbidden by law to contact us. If this is simply some other alien unassociated with the Zom, which I doubt, then he's either an agent from a rival alien government, or some kind of renegade."

"Why don't we run the face and see if we can connect him?"

"That's just what I was thinking, but that doesn't solve all of our problems." He stood and began to pace the floor. "Rod, up to now I've been tolerating Stan's little project, because it has always irked me the way we laid down before the green guys so easily; it just seemed so...un-American. There we were, on the verge of some very big things, and then suddenly we were expected to just look down at the dirt and pretend as if space just...isn't there. I have always been a fan of space exploration and I think that what Stan is doing is a good thing. As long as the little green guys don't know about it, I am all for it. Now it looks like maybe they do know. I can think of no other reason why any alien, Zom or otherwise, would be talking to our friend Mr. McPherson. The NSA and the IRS are both breathing down Stan's neck and he knows it. E.T. here is probably making an offer to Stan, to help him finish his work, squeezing him to get him to spill the beans—that's how they operate. If Stan ends up trusting this fellow and he turns out to be a Zom agent, not only could it cost Stan's his life, but you and I will be left answering some very difficult questions about what we knew and when we knew it."

He sat back down and shook his head.

"This cat may be out of the bag, and if so, we need to plan some damage control. It may be necessary for us to pull the plug on him before somebody else does. At least if we do that, he's more likely to come out of it still breathing."

"We'll still have to find where he's building it."

"Ya, but that shouldn't be too difficult, once this stops being our little secret and we put the whole department on it." He thought for a minute, "Let's keep watching this and see how things develop. Stan is secretive, clever, and mean; this might just blow over." He pointed to the photo, "and see if you can find out who this green guy is and what he's up to."

"Ok," Mills stood up and left, taking the photo with him.

He had some calls to make.

Bill Housley

Chapter 3

"Your designs are sound," the alien said, as he looked over Stan's research. "I am not an engineer, but it all looks good—primitive but functional. We might even be able to help you improve it in one or two little ways." He looked up and smiled, "It might be easier if I could inspect the vessel itself, is it nearby?"

Stan had let the alien, Clell; spend the night at his estate so that he could watch him. After breakfast, they sat in Stan's study and worked over coffee.

"I'm not sure that I'm ready to share that information with you just yet, Clell. You seem to think you can just walk up to me and say, 'Hi, I am from outer space, tell me everything!' I have friends who have a lot deeper history with me, who know little or nothing about this project." He folded his hands on his desk. "You expect me to trust you, but trust starts with mutual understanding."

"Those circumstances that I spoke of yesterday are desperately urgent," Clell said, leveling eyes with him. "There may not be much time."

"I need to know some more about you and your people." He took out a pencil and paper. "Those 'circumstances' you mentioned seem like a good place to start."

"Well," Clell said, "I suppose I have been pushing you rather fast…and you are right, good relations are grounded in good communications." He sat down across from Stan's desk, took the pencil and paper away from him, and drew as he spoke. He turned an empty coffee cup upside-down on the paper and used it to trace a large circle. Then he used the same cup, turned right side up, to draw another circle about half that size right next to it. Then he drew a larger, ameba shaped body encompassing them both. He finished by placing a dot inside the larger circle, close to the point where the two circles nearly touched.

"I come from a world that is a member of what we call a star-nation, which is a group of stars and their planets, administered under a single government. It, in turn, is a member of a large consortium of planets and star-nations, very loosely governed by a bureaucracy called the 'United Council for Peace', which can best be described, I suppose, as a cross between Earth's 'United Nations' and a commonwealth. It was formed several hundred of your years ago."

He pointed to the dot. "This map is an over-simplification, but your world is about here, within the borders of what is called the 'Zom Republic'. This other circle is another star nation called the 'Crenum Alliance'. They don't always get along very well and tensions between the two arise from time to time."

"This nation, the Zom, want to setup some mining facilities on your planet. The Zom economy needs this mining very badly, since the resources in question are critical for vessels traveling from place to place, as well as the many worlds and outposts in the Republic that are not located on what we call 'living' planets like this one."

"What kind of 'mining'?"

"Biological consumables."

"You mean food?"

"Yes, foods, certain medicines—various other supplies necessary for sustaining life—many such items cannot be synthesized efficiently, but grow in natural abundance on living planets. Our economies mostly obtain such things from the more primitive worlds—systems with larger proportions of their populations living in orbit, and on dead orbitals in-system, like dead planets and moons and things, end up with a smaller surplus of biologicals that they can export."

"Makes sense."

"One difficulty they have is that the UCP has laws that forbid interfering with developing, grounded cultures."

"Grounded?"

"It means people without light-speed travel technology. Many nations in the UCP and the Crenum Alliance in particular, consider your world to be culturally valuable and want the Zom to leave it alone until it matures. They feel so strongly about this point that they are willing to go to war over it. It is a big issue for us. If allowed to develop properly, your world, positioned near the border like this, could eventually petition for independent status under the UCP charter, maybe even become a trading partner of the Crenum Alliance instead of the Zom.

"You see, there was an incident some time ago, called the Mistaac Debacle. The Mistaac were a very warlike and aggressive people that several other worlds nearby had to put down violently for their own self-defense. The Mistaac were defeated so completely that the Mistaac race could no longer sustain itself and died out. Afterward, an investigation into the history of the Mistaac found a lot of contact with alien technologies had accelerated the development of their sciences, causing the more short sighted and violent elements of their culture to overcome the peaceful, progressive ones, forever altering their future and making them a threat to the region. They sought to subjugate their neighbors, and when their neighbors fought back they wiped them out, not realizing how thin their strength was.

“After that, the UCP passed a non-interference pact, called the 'Mistaac Accord', protecting developing worlds from advanced technology until they get to a point, on their own, where they can reach so far out into space that contact is with other cultures is inevitable anyway. At that point, they might be given independent access to representation in the UCP, trade status, UCP member currency, that sort of thing. Then they can choose their own direction in our world.”

Stan pondered this. The Zom, whoever they were, needed packaged food for their people who lived aboard starships and on barren worlds—that much made sense—but why would they be willing to provoke a war just for one food source planet? For that matter, why would the Crenum want to start a war against a larger adversary, for just one more possible trade partner? It didn't add up. Why would the Zom, or the Crenum, risk the expense, devastation, and loss of life in a war, fighting over Earth? Unless…

“That can't be the only reason, Clell. I mean, the way I see it, with Earth sitting right between the Zom and the Crenum like this, it looks to me to be better positioned as either a trading post in times of peace, or a jumping off point in times of war. If we were further in, we would make a good resource planet, but here we look—to me at least—like a nice place to house and feed a garrison.”

Clell looked at him for a moment and then nodded, “Yes…there are some Crenum who are saying that, because there are five important systems near here that are Crenum, not Zom. In a time of war, the Crenum would have to post strong enough forces at each of those systems to defend any of them from attack from any forces stationed here.”

“So, Clell, you told me what could happen if we developed into one of your equals, or if we are *developing*. What if we stayed just as we are now…pretty much…stuck? What would happen then?

Clell paused.

"Well, I have been studying some of your history, so I will use an example from there. Do you remember the cultures that dwelt here on this continent, before the Spanish came?"

"Yes, of course."

"They were pretty much technologically…'stuck'…as you put it."

"Yes, they were."

"Your people's fate would end up being something like theirs."

Stan felt a chill.

"That would destroy our culture."

"Your culture would be the least of the losses, if it sparks a war," Clell said as he leaned back in his chair. "Interstellar cultures have far too much access to one and other, with weapons that are far too powerful. For us, even a war between small participants can escalate into an Armageddon for one or both of them very easily. Both the Zom and the Crenum are very large, influential, and dangerous nations and there is no telling how long a war between them would rage or how far it would spread. Some have predicted a complete, catastrophic collapse; the dissolution of the UCP and the loss of many ancient and advanced civilizations. The potential damage is immeasurable."

"I can appreciate that, believe me. So, applying your metaphor, if we are judged as static, not progressing, the Zom might…invade and enslave us."

"Would…would invade and enslave you. The Zom already consider you as 'static' and are putting together their plans even as we speak."

"But your laws…"

"Would offer you no protection at all—our laws do not protect static worlds."

"What if we weren't quite so 'static'?"

"It is believed that the Zom will not go through with their plans if they are not on sound legal footing," Clell sighed. "It is a complicated issue of perceptions, alliances and so forth. If other worlds felt like the Zom broke the law, then more of them would come to the aid of the Crenum, but if the Zom were viewed as legally within their rights, then the whole strategic picture changes and puts them at a severe advantage."

Stan thought about that for a minute. Wars were always justified, by both sides, before they began. Could it be that this was a war that was already coming—that all of this was just the usual posturing and positioning that came before the storm? Could such concepts be so universal? For that matter, why had Clell not spoken of the obvious military possibilities of Earth's location—until Stan mentioned them?

"Which side are you on?"

"Me?" Clell picked up a stack of Stan's printed notes and thumbed through them as he thought for a moment. "I am an agent for the UCP. I am here, covertly, to see for myself if this ship of yours really exists, and gather evidence that could qualify your world for protected status, as a developing culture, rather than the failed and declining society that the Zom seem to think you are."

"This planet is *not* ready for any kind of other-world alliances, trade agreements, or membership in that UCP you spoke of," Stan shook his head, "and the existence of my ship does not change that."

"I agree with you, but the Mistaac Accord can still protect it until it is. The problem is that, right now, your world appears static. Many planets pass their peak without ever developing interstellar travel, and then they never get there at all. Some races in the UCP regard the people of such worlds to be inferior—fit only for slavery and exploitation—much like your people regarded stone-age savages hundreds of years ago. Many claim that badly needed resources would end up sitting on those planets, un-used. So there were exemptions written into the Mistaac Accord which disqualified those worlds as…unworthy of protection. Your ship is the hinge point. It could be seen as a sign of progress, and the law could continue to shield you from alien contact."

"I can see why you need to be covert about being here too," Stan pointed at him, "You have been sitting here looking at my data and discussing it with me, offering to help me, in a way telling me that I'm on the right track. So aren't you breaking your Mistaac Accord? In a way you are practically collaborating with me already."

"I have been breaking the Mistaac Accord from the moment I approached you yesterday evening," answered Clell, leveling his tone. "Not to mention violating the sovereignty of your country, and of the Zom Republic. If I am caught here by any of those governments it would trigger a very large and unpleasant scandal." He started to grow angry. "In fact, I am here at very great risk to myself and to those who sent me, and every minute that I spend, sitting here in your house, jeopardizes everything that I am here to do and brings us closer to the invasion of your planet by the Zom. Stan, if you only knew how quickly they would dismantle your military forces. It is far worse than anything either your meager history or your misguided science fiction has prepared you for." He stood and pounded the table with his fist. "These delays are stupid! The time for talk is over! You must stop wasting time and tell me the location of your ship!"

Stan sat blinking for a moment, shocked at his outburst, and then stood up.

"Those things which you call 'meager history' and 'misguided science fiction' form important windows into my culture. They show us to be a resourceful and very violent race. If you really knew what was contained in our history and literary works, you would not be so convinced that we could be steamrolled easily, technology notwithstanding. I am not saying that we could win, necessarily, but we might not lose all that quickly either and you look like your kind bleeds just as easily as mine does."

They stared at one and other for a moment. Then Stan took a deep breath.

"I will not tell you where my ship is hidden," he continued quietly. "You'll know its location soon enough. I'll take you there tomorrow. Until then, you will not mention it again. If you do, I will report your presence to the Earth authorities. We are not so primitive anymore that a complaint from me, to the right people, will not be believed. We are uncivilized enough, however, that your people might never see, nor hear from you again, except to read about the size and shape of your internal organs in some declassified government document."

Clell sat down, leaned back in the chair, and propped his feet up on the desk.

"Very well," he said, "I've heard of a game your people play called 'Chess', would you mind teaching it to me?"

***

"Ambassador Fairdock! Good to see you."

The Crenum Chief of Interstellar affairs, Roche Gimlar, stood and offered his warmest welcome to the Zom ambassador to the Crenum Alliance, Mer Fairdock.

"Likewise," replied the Zom emissary. They each bowed twice to the other, first to the left, then to the right, the traditional Crenum greeting, reserved for formal occasions and semi-formal meetings such as this.

"I was expecting your message, and came right over as soon as I received it." Fairdock said as they sat down.

Gimlar took a breath and began, "I assume then, that you've read the transcript of the latest UCP session?"

"I have, and I might say that it is most unfortunate that our delegates at the UCP have such a contentious air, it seems they're rather caught up in the hysteria of late—not like you and I."

"No, not like us." *You little worm, you hate my guts as much as I hate yours, and we both know it, but I'll play our little game.* "I've invited you here today, so I could explain some of the hasty words of my colleague, Ambassador Klicktah. In his speech, he made some unprepared and unapproved statements in which he alluded to some things that are not actually in evidence—and rather thinly sourced. We Crenum pride ourselves in accuracy. It was, of course, somewhat embarrassing."

Fairdock smiled and nodded.

"Well, he did leave quite a strong impression that he could support this...claim of his...regarding a Telesian spacecraft," Fairdock said. "He ruffled quite a few feathers in my government, and embarrassed the administrator of that region...but after consulting with him; we knew that such a project would be quite impossible. I doubted the accuracy of the report anyway, because it came from the Crenum UCP delegate instead of from you. Such surprises are normally aired here in your office, not in public."

"Yes, of course. I spoke with Klicktah and he regrets his comments. Nevertheless, I knew that you would want the matter investigated internally, so I sent out some inquiries. I spoke with the chief of our intelligence services; he sat right where you are sitting now...he denied everything of course."

"Of course. I hope you looked deeper."

"I did...I still have some influence within our intelligence community. After digging around a bit I managed to stir up a rather interesting document. It contains some details that may have formed the basis of Klicktah's claim regarding this alleged Telesian starship."

"Indeed," Fairdock said with interest.

Even though he disliked the man, Gimlar enjoyed working with Mer Fairdock, and had a very high professional respect for him. He watched Fairdock's eyes carefully and saw nothing; the Zom had chosen their ambassador well.

"Yes, and in the interest of good relations with your people, I managed to twist some ears and persuade them to let you view it."

Gimlar handed a slip of paper to Fairdock with a flourish, hoping that he wasn't overplaying it. The document was a complete fabrication—an obvious forgery—and Fairdock would know it. Gimlar had finished drafting it only moments before Fairdock walked into his office—the ink was barely dry. It purported to be a memo from a Crenum field agent—who did not exist—who claimed to have come across news of a spaceship under construction on Telis 3. It also claimed that the source of the information was a communications intercept from some interstellar race operating on the planet, though the memo failed to identify whom, but very specifically named certain high officials within the government and intelligence services of a country called "The United States of America", on Telis 3. The document had some blacked out portions—to leave the impression that it had been "cleansed" of things like the agent's code names, locations, and other operational data. The Telesian officials that Gimlar had named in the document were, however, quite real and he had taken great pains to spell their names correctly, even to the extent of using properly detailed Telesian symbols. He had even obtained some authentic Telesian documents, to help him to render their written language properly.

The fact is, Gimlar knew from some other reports he had access to, that the very same Telesian officials named in his fake memo had been working closely with the Zom for years. His data, gathered from multiple and very reliable sources, said that Mr. Fairdock knew who those officials were. Fairdock himself, on a previous job, had even met some of them.

Gimlar watched Fairdock carefully as the ambassador read the memo, marveling again at the ambassador's self-control. Whatever anti-perspiration medication the Zom ambassador was taking was now demonstrating its potency. All clandestine contact with members of a grounded culture, like the one on Telis 3, was strictly illegal under the Mistaac Accord. Such contact might have far-reaching repercussions if it ever went public. If government-to-government contact was necessary, the UCP preferred that its own officials do it, very carefully, and only for very good reasons. Gimlar had begun to accumulate significant evidence that the Zom had been flouting the law with regard to Telis 3 for many years. Their conduct had been abominable, to the extent of frequent, routine, and sometimes even violent contact with its people, and outright meddling within some of its governments. Their interference had been so disgracefully direct, and desperate, that Gimlar doubted in his heart that the Telesians' current stagnant condition was entirely the Telesians' own doing, though he couldn't prove it yet.

"Now Mr. Ambassador, I assure you that I am pulling out all of the stops on this matter. I promise I will get to the bottom of it, and if any of our people are involved in this...obviously illegal...interference with the Telesians, I will put a stop to it immediately and see that all of those involved are swiftly and thoroughly punished."

Fairdock tore his eyes away from the names on the document, "I...I would appreciate that, Mr. Gimlar, such a violation of our sovereignty and of interstellar law would be most grievous if the public were to get wind of it. I hope that the information on this document has been thoroughly...contained?"

"Oh, it has…for now. I am only showing it to you so we can work together on it. There are certain events unfolding on your side of the void…events that…if they develop too soon…could drag the details of this matter into the public eye before you and I have had a chance to…package them appropriately. I was wondering if you could ask your government to delay any new projects in the area of Telesia that might attract…attention…until you and I have had a little time to work this issue through."

Fairdock seemed to ponder for a moment, then smiled thinly.

"I see your point, and I am glad that we agree again, Chief Gimlar." He stood and bowed, leaving the document on the table.

"As usual, I come away from this meeting with strong confidence in a continued peace between our two peoples."

"As do I." Gimlar allowed himself a savage grin as he watched the Zom ambassador turn and head swiftly for the door. Blackmailing such slime as the Zom was an activity he *should* relish, not regret. He hadn't stopped the Zom's invasion plans, he knew that, but now they would have to clean up a few loose ends first, disconnect themselves from any evidence of impropriety that they thought he might have secured, before they could proceed. It wouldn't take them very long, but maybe that, and Klicktah's activities, will delay them long enough for Gimlar to get his hands on that spacecraft…

…and save Telesia.

***

During the thirty minutes by private plane, then two hours by four-wheel drive over kidney-busting dirt roads through the Sierra Nevada, Stan had some time to think.

Few skilled liars ventured very far from the truth, and if any of the background information that Clell had given him was even remotely true, then the situation had just become very desperate—and dangerous. Stan had let Clell fly the plane, partly because the alien was interested in the primitive controls, but also because Stan wanted to keep him busy so he could observe Clell with his guard down.

Stan had noticed that Clell occasionally displayed arrogance and impatience toward Earthlings, a trait which might be a common trend among aliens of every stripe, but it was a tool that Stan could use. It sometimes caused Clell to slip up and speak a little too carelessly, especially when he grew distracted by anger or ease. Stan paid close attention to everything Clell said at those times, and remembered every detail.

Most lies are merely half-truths, correct facts manipulated and distorted to support the liar's aims. Stan had already surmised that the story Clell had told him the day before was neither complete, nor was it completely truthful. There were a couple of small clues that Clell had let slip; in particular, Clell had looked away before telling Stan about being employed by the UCP. Stan had sprung that question on purpose, and specifically watched for Clell's reaction.

The similarity of alien physical makeup to humans interested Stan as well. It seemed incredibly extensive, and it was possible, though unlikely, that Clell might simply be a human agent in makeup. But whether lying human or lying alien, Stan enjoyed having the consistent clues to work with.

Clell's claim that time was critical was obviously true. Whoever or whatever Clell was, he was walking, talking proof that Stan had attracted too much attention. If it was true that Clell was there to help, then he could be a valuable ally, but if it turned out that he was there to compromise Stan's project, then Stan had to be ready for anything.

Some in the intelligence community already knew that Stan was building a ship—no one knew where. The trouble involved with finding the ship, and the lack of general surety of its existence, had so far kept his project safe—that and the fact that those few who knew about it secretly wanted it to succeed. Only Stan knew how close he was to completion, and exactly what his new creation was capable of doing once it was finished. The mathematics that had been worked out during the PAGA drive project at NASA had shown that a spacecraft so equipped could accelerate by as much as one Astronomical unit (the distance from the Earth to the sun) per day, per day, but there had been flaws in those calculations. Once he had had actual flying models to work with, he was shocked at how inaccurate those early formulas had been. His first test probe, built shortly after NASA had cancelled the project, made the round trip to Mars and back in a few hours, and then later traveled to Alpha Centari in a little over four weeks, roughly half a light year every 24 hours. His newer probes had pushed the envelope even further. Their acceleration rates and maneuverability, even within the atmosphere, had been nothing short of astonishing. Once the ship was complete, it could probably escape pursuit by almost anything man-made. He had intended to take the ship off planet eventually anyway, but now that it appeared that off-worlders were involved, Stan couldn't allow himself to be sure of anything.

He knew of several remote and desolate places on Earth where he could go; getting into space would have to wait. He had to get back to work, now, and keep working non-stop until the ship could at least fly. Clell would probably follow him everywhere, whether he wanted him to or not.

He had taken certain precautions against treachery, however. He felt a little bit guilty for his duplicity, but it was prudent to take some security measures. He had secretly armed himself. It was only a derringer, hidden in his pocket, but it was better than nothing. Stan liked coming across as a helpless, harmless scientist, and guessed that Clell was as vulnerable to "primitive" firearms as humans were—though he hoped he wouldn't have experiment on that.

The builders of the artificial tunnel had bored straight through the mountain and lined it with concrete and steel. He preferred to use this back entrance so that the main entrance around on the other side of the facility would continue to look deserted. The road there was paved, but cracked and over-grown, and he only used it when he had to bring in a large, late-night shipment by truck.

He looked over at Clell, but the alien was busy examining the mountainside and telling Stan how the arid plant life growing there was similar to some of the species on his own home world. Sometimes this Clell fellow seemed almost human. His grasp of English, as well as his knowledge of and interest in manners, customs, flora and fauna were certainly impressive.

Maybe I am just being paranoid, Stan thought.

Then his watch vibrated; someone was sending him a message. His wristwatch was an advanced computer that he routinely used to access email, the Internet, his ship, and all of his other computers when he was away.

He glanced over at Clell again, and then looked at his watch, as if to check the time. There was a text message on it.

Your new friend
is dangerous
and he is not on
your side of things.
Gastone

Stan's blood turned to ice. "Gastone" was the code name that his CIA friend used whenever he made contact. The man had worked with him on a couple of side projects involving his ship, including his little midnight burglary of Kennedy Space Center to steal the PAGA project data. "Gastone", aka Rod Mills, sometimes sent him carefully coded messages, but Mills was almost never the first to make contact, and they hadn't communicated at all for some time because of the increasing threat to both Rod's position at the agency and to Stan's illegal project.

This message, using such comparatively open terms, could only be urgent. Mills was a prudent man, not likely to take risky action without good cause. This email, should it be intercepted, could be easily understood by someone else—he had apparently wanted Stan to get the message quick and clear. Rod worked at the CIA, and was loyal to the agency, but he also cared about Stan's welfare. On that score Stan trusted him far more than he trusted this mysterious alien.

He had not been aware that the agency was watching him so closely, but how Mills knew that Stan even had this "new friend" was not as shocking as the fact that Mills knew who that person was. His imagination reeled at the various permutations. It put in doubt many widely held assumptions about his government's focus and replaced them with kook theories from the lunatic fringe. Mills and the CIA seemed to know of alien agents operating on Earth—knew them deeply enough, apparently, to know the specific characteristics and motivations of this particular one. Combining that with Clell's revelations from the previous day gave Stan a headache.

"Dangerous" and "not on your side" were not expressions that Stan liked to hear about a person who was about to learn the exact location of his grounded ship. Yet here they were, less than a hundred yards from it. He couldn't just leave; Clell was close enough that he could come back again later and find it on his own. For Stan to turn unfriendly on him was even worse, because it could end with his death.

What a fool I am! Stan thought. *Now what do I do with him?*

The more he thought about it, the harder it was to find a solution that he liked. If he had gotten the warning earlier, he could have just changed his mind and brought Clell back to his home to talk some more, or just lead him somewhere else and ditched him. Now he was in a corner. Every plan his brain went over seemed to end only one way. If Clell was supposed to take some kind of action once he found the ship, then he might realize at any moment that he was close enough to it now that Stan was expendable. It was a time for quick action, and the only action Stan could think of was murder.

"Looks like we've arrived," said Clell with a cheerful smile. "I can't wait to see Earth's first starship. Let's go on in before my head explodes."

Stan got out of the jeep and dug both hands into his pants pockets, fishing for his other keys. It was dark in the cave entrance, dark enough, Stan hoped, that Clell wouldn't see the derringer that he palmed in his left hand when he pulled it from his pocket. He unlocked the door and pushed it open, standing aside as Clell drove the jeep in, then closing the door behind them. Stan deliberately refrained from turning on the light, prompting Clell to switch on the jeep's headlights. Stan risked a glance at the alien, who was watching Stan intently and seemed a little bit annoyed to be looking at another set of doors. The entryway had been built as an airlock, to keep the inside of the facility clear of dust. Stan used the enclosed space in between as a parking garage whenever he came. Today, it would serve to keep Clell away from the ship a minute or two more.

Here it comes; if I don't do this right, I'm dead, Stan thought to himself. *I'll get behind him while he's looking at the ship and disable him; then I'll have time to figure out what to do with him.*

Clell parked the jeep and got out, looking around the small entranceway carefully. The two of them walked across to the inside doors, with Clell making an obvious effort to keep Stan in front. Once at the door, Stan reached across with a shaking right hand and touched his biometric control panel. There was a beep as the laser inside scanned his handprint and the door slid open with a hiss…

…and Stan found himself looking into the masked and helmeted faces of ten people…

…and ten drawn weapons.

Chapter 4

Stan turned and looked back at Clell, and right into another weapon. Standing in the doorway, he had nowhere to go.

Clell said something to the others. He spoke in a language that Stan didn't understand, but he definitely did not like the way he said it. Clell's voice hissed with menace.

One of the alien soldiers stepped forward and answered back calmly, in English.

"You lack imagination Clell. You think that killing something is the solution to every problem."

Clell shouted back at him.

"You are being rude Clell. Speak his language—I know you know how."

Bewildered, Stan pondered the situation as his heart raced. Standing between them, he turned sideways so that he could look back and forth at each. All weapons seemed trained on him, or past him, Stan wasn't sure which. At least he was sure they were weapons. He had never seen the devices before, but the chill that crawled up his spine told him what they were. He still palmed his cocked derringer, but what good was it against eleven targets?

Clell spoke again, with a nasty grin on his face. Slowly, the others lowered their weapons a little, but Clell's gun was still pointing at Stan.

All of this was taking place in the very shadow of his uncompleted ship, there in a cave, far beneath the notice of the world. Whatever happened next, no one would help. No one on Earth would know. Who were these new players? Had Clell already betrayed him to these other people? One thing was certain; Clell looked less like an ally with each passing moment.

The person who had been speaking to Clell holstered his sidearm and removed his helmet, showing a smiling face underneath, but his eyes looked hard and worried. He was, like Clell, nearly human in his features, but different enough from Clell so that Stan guessed he was from another alien world entirely. He bore a different shaped head and eyes, and a slightly different complexion.

Clell motioned with his weapon a little, and his reply was cold and final, spoken through clenched teeth.

“We're walking out of here, the Telesian and I. If you try to stun me, I'll kill him.”

“No,” said the alien leader. His tone had a nasty finality of its own, and his eyes were grim. "We need him, and if you kill him you will never make it to the outer door.”

Clell motioned with his weapon to Stan, “Come along and don't argue. This can stop your heart instantly; it would look to your people like a simple heart attack.

“And then we will arrange a similar little accident for you, Clell,” the other alien said, with a deadly grin. "This mountain is a big place, full of dangerous things that can hurt people.”

Clell spoke something angrily in their alien language again.

As they argued, Stan felt like a mouse, crowded between two cats. He put his left hand on his thigh, worried that the derringer might slip from his sweating palm, and thanked the darkness.

What to do? Stan pondered.

The other aliens joined in the dispute, and the tension mounted with the volume. It was obvious that Clell and these others didn't like each other very much. They appeared to be rivals, maybe even enemies—though he recognized, that in this particular situation, the enemy of his enemy wasn't necessarily his friend.

Stan's life, his ship, and perhaps much more teetered on a precipice of disaster. These new aliens held the numbers, the ship, and apparently some regard for his safety. Enemies or not, there seemed to be nothing he could do about them. They were very at ease, almost at home in their confidence, and he decided to treat them as friends—for the time being at least. He was sure, now, that Clell had been a dangerous enemy all along.

He had to do something, but Clell's weapon still pointed at his chest and he doubted he could raise the derringer and fire first. Clell could not shoot these other aliens because they were too many, and they could not shoot him because Stan stood in the way.

The seconds ticked by in minutes, it was maddening. They debated in hot tones, deciding the future of his ship—his future—and perhaps that of the whole planet. He deserved to have a role. He decided to take a chance.

"So...Clell, help me out here. You aren't *really* some kind of special agent for the UCP are you?"

The arguing stopped, the topic suddenly forgotten, as the alien leader and several of those with him laughed.

"Is that what you told him? What other wild tales have you been filling his head with, Clell?" the leader asked. He shook his head. "No, Mr. McPherson...Stan, Clell is not any kind of UCP intelligence operative, there is no such creature. The UCP has no intelligence agency."

Clell snorted with derision.

"You and these other folks then, are...Crenums?"

"Correct Mr. McPherson," said the other alien. "We are from the Crenum Alliance. You may not know very much about us, but we have followed your work for a long time. I can tell you more later. I hope you understand by now that this Zom here cannot be trusted."

Stan looked back over at Clell.

"Clell has already educated me somewhat concerning you. I started out thinking he was a friend, but I'm getting wiser." He gave Clell a hard look.

"You lie," Clell sneered. "You are a gullible fool. You have known these others for only a couple of minutes and you trust *them* already, just as you started out trusting *me*."

"I did trust you, but not completely. Your mission all along was to betray me and my ship, wasn't it?"

"More or less."

"You were very informative last night, thank you. You spoke of the UCP as if it respected the sovereignty and intelligence of primitives, yet your own attitudes seemed different."

Hold it together Stan, he thought to himself.

Clell ignored Stan's comment with a sneer.

"It doesn't matter now, it is plain from the look of that vessel that it is not going anywhere soon, these Crenum spies cannot harm me without risking a war, and you are helpless to do anything either way. My work here is done—I will not need to hurt you. Come with me, now, my people will be along later to deal with these trespassers." He waved his weapon slightly and stepped a little to one side.

"Not so fast. When you call them trespassers, you say it as if you own the place. The way I see it, you're all trespassers." Stan's palms were sweating even more now, and his legs and voice started to shake. He tried not to glance at Clell's weapon, willing him to point it somewhere else, if only for a moment. Stan had practiced a lot when he was younger and had been a half-decent shot once, but at such close range he wondered if accuracy really mattered anyway. He went through the movements in his mind, resolving not to accidentally fire early, nor to wait too long.

"Earth may be in Zom space," he said, struggling to calm his voice, "but you're still from another planet, Clell. You don't belong here any more than they do. You've inserted yourself into my life, told me of some of your technology…violated your laws. That…" he snapped his fingers, "what did you call it…Mistaac Accord? Am I right? Is there such a statute?" He looked at the Crenum leader.

"There is," he nodded, looking quizzically at Clell.

"Is it true, then, that were it not for my ship his people would be basically free to invade?"

"There are some parts of the law that are somewhat…ambiguous," the alien said, "and I disagree with the way the Zom prefer to interpret it. Essentially, if good, clear evidence of your ship were to become widespread among my people, it would make the matter very much clearer." He leveled a steely look at Clell as he spoke.

"Earth is a backward, dysfunctional…primitive culture!" Clell spat. "Your ship does not change that, you told me that yourself! Now come on, Stan, we are leaving. Do not make me tell you again!"

"Do not bully me, Clell," he was trying to act brave, but his guts were ice. He had faced fear and death before, had seen combat briefly as a fighter pilot, and rode a rocket into space—he didn't consider himself a cowardly man. It was different this time. Back then it was his job to be brave, and if he died he was a hero. This time there was anger—anger mixed with *real* fear. Fear not only of death, but of failure, of insignificance—fear of dying in a meaningless way and never making a difference. Fear of complete and permanent failure. He had never experienced *that* brand of fear before.

"These others may be spies to you, but they are also witnesses," his voice was growing shrill. "They have already seen this ship, probably photographed it. You can take me out of here—and I don't believe you can afford not to kill me—but even if you call your people right now, I'm guessing they are not so close by. These others will escape and report what has happened here." He was grabbing at straws now, while his heart rattled around in his chest like a caged animal.

"Besides, you people seem to think that this is all about you," he still watched Clell's weapon out of the corner of his eye. "If your people and your disagreements didn't exist I'd still, slowly, complete this ship without you. In fact, it seems to me that you have issues to settle between yourselves!"

Stan continued loudly, feeling bolder. His own words gave him an idea, he only hoped he could hold his nerve long enough to pull it off. "Now that I've met space aliens, you really are not all that strange; nothing like what we imagined. We've always believed alien species' would have it all together, be more civilized than us somehow, but if you people looked any more human no one would know you weren't from Earth—the way you behave."

Clell and the alien leader looked at each other, apparently unsure where Stan's outburst was leading.

"You cannot seem to discuss your differences like civilized beings, anymore than we Earthlings...or Telesians or whatever." Stan said this with a wave of his empty right hand. "You do, claim, to be civilized right? You say we're primitive, Clell, but none of you are acting particularly *advanced* at the moment, and this Earthling, for one, is not impressed!"

"My people and I are here to help and protect you," the Crenum leader said calmly, "and I suggest..."

"That's what this liar said!" Stan shouted, pointing a finger in Clell's face, almost touching his nose. Clell didn't recoil, but just stood there with a half-smile on his face. Stan looked back around at the others. "How do I know you're not lying too?" With the back of his head to Clell, he winked before continuing. "Up to now, I've worked on this ship alone, making slow but steady progress. I didn't need you then, yet ever since I started accepting 'help' from you so called 'advanced' life forms everything in my life has gone straight to Hell! Some help! I'm better off without you! I see now the wisdom of your law...that...Mistaac Accord of yours. Sounds like someone enacted it to keep you from screwing things up! It seems like a very wise law, and even though I am not bound by your laws, I will help you obey them. Get out!" he said, waving his hand as if to shoo them all away. "Off my property! Clell, you too! All of you!"

The Crenum leader could not have missed the wink, and seemed to think for a moment.

"You are making a fatal mistake my friend..."

Stan held up his left hand toward the alien, as if to interrupt him—it was the one palming the derringer.

"My mistake, *friend*, was in not telling Clell here, two days ago, what I tell all of you aliens now. I have a plan. I'm doing fine. I don't need you. Thank you. Go away!"

The only Crenum who had removed his helmet was the leader, and his eyes flicked quickly to the tiny pistol, hiding in Stan's outward facing palm, and then back to Stan.

“You are right,” he said, suddenly growing angry. ”I do not know why we even bothered wasting our time, and risking our lives, coming here. You know, I have been on your planet, studying your culture, and I have found Telesians to be a primitive, self-absorbed, corrupt, short sighted, arrogant child-race. We came here thinking *you* might be different, but I see now that we were mistaken.” He said something in his own language over his shoulder and his people all holstered their weapons and began to disperse. ”We will leave, and I will recommend to the Crenum government that we ignore you. You are hopeless. You and your people really are ignorant savages and I leave you all to burn in the fire you have kindled for yourselves! Enjoy it! Goodbye!”

He looked past Stan, “As for you, Clell, this primitive is right about one thing. We *will* be reporting on your activities, and we have more to report than just this business here. I promise you, Clell, if it takes me the rest of my life, I will testify against you at a UCP tribunal one day. You and your people will pay for your interference with this planet's development.”

He turned to leave and then stopped, as if suddenly thinking of something else.

“Bury him deep, Clell,” he said over his shoulder. ”The scavengers around here are large, and will dig up his remains.”

Clell lowered his weapon, but showed no intention of going anywhere.

Stan felt a fresh rush of adrenaline. The time had come.

He must protect his ship at all costs. Clell himself had told him—and these aliens had confirmed—that this spacecraft was more important than Stan had ever realized. These Crenum, at this moment, did not appear to threaten it, but Clell did. Billions of lives were at stake—and not just Earth lives either. How had he gotten himself so caught up in affairs so far from home? It had never been his intent. He worked for himself, for his country, and for his world. He wished things could just go back to normal. Somehow, he knew they never would, but at least he could try to bring circumstances back under his control.

Some of the facts still seemed fuzzy, but he understood now that everyone in that cave had many reasons to kill to protect their interests…

…even a harmless, helpless, naive scientist like Stan McPherson.

His head and face felt hot, it spread down his neck to his shoulders, and his hands began to shake. There was a ringing in his ears, mixed with the sound of his own heartbeat. He released the derringer, letting the handle fall into his outstretched fingers. Then in one movement, he lifted it up, pointed it in the direction of Clell's head, and fired. The gunshot sounded far away as the tiny pistol jumped in his hand and the sting of hot gunpowder hit his face. Clell faltered backward and fell.

Just then, some kind of force passed through Stan from behind. His body suddenly felt limp, and his world spun around him.

Then everything went dark.

***

He awoke in his cot, onboard his ship, in the temporary living space he'd setup for himself inside the cargo hold. The aliens seemed to have taken up residence there as well, and the room was a bustle of activity. Some sat or stood at various pieces of equipment that Stan recognized as his, while others operated tools and gadgets that he'd never seen before.

They were wearing different clothes, and the helmets and most of the weapons were gone. Stan started to get up.

"Stay still for a while longer," he heard someone say. He felt a firm but gentle hand on his shoulder. He looked up into the face of a Crenum with kind and laughing eyes.

"What..." Stan began.

"Do not worry," the man interrupted, "Colonel Grandin has been waiting for you to wake up, and he will answer all of your questions." He picked up a walkie-talkie from nearby, one of those cheap, short-range ones that are available at practically any electronics store on Earth, and spoke into it using the same alien language that Stan had heard earlier. After talking over the radio to several others, the man nodded with satisfaction and turned back to Stan.

"The Colonel has to leave soon to perform some other duties, but he wants to talk to you first."

"You use Earth technology?"

The man looked puzzled for a moment at the question, and then looked at the walkie-talkie, turning it over in his hand. "Yes. Sometimes we do, when it suites us. We try not to expose your world to our toys if we can help it. Besides, these devices are handy for security purposes. They use radio waves that don't penetrate the outer walls of your facility."

"Of course, I should have realized that. You are a medic?"

The man nodded.

"What happened to me?"

"You were shot...stunned really. Grandin should be the one to explain that, I think," he smiled again. "The effects are only temporary."

"Sir, the man I shot..." Stan wanted to explain to someone why he had done it; he already felt guilt over it.

"...he will live, and will recover fully," the medic said with a mischievous smile, "though it would have been much simpler for us if your aim had been better. Anyway, perhaps the small area of Clell's brain that you damaged is the part that makes him such a Gerp."

Stan felt relieved. Just because he had decided to kill, didn't mean he liked it. Part of him was glad Clell had lived, another part wondered if he hadn't just left a snake loose in the bed.

"Good, you are awake!" Stan heard a familiar voice and looked up; it was the alien leader from the cave entrance. "My name is Colonel Glem Grandin. Sorry I shot you, I couldn't be sure you had completely disabled Clell, and I couldn't stun him without hitting you. What you did was very brave, and I could not afford the risk that he might kill you for it—it would have been a great loss." He stuck out his hand and helped Stan to a sitting position; then sat in a chair by the side of the cot. He moved like a man accustomed to taking action, strong and quick of both muscle and brain. His eyes measured Stan as he talked with him and his voice had the tone of command. Stan liked him, though he questioned the man's definition of 'brave'. He thought what he had done had been more desperation than bravery.

"I assume that you did not mean what you said earlier about wishing us to leave. We have a lot of important work to do here, but if you are serious about it then we really would have to go."

Stan thought, though his head was still spinning a little. Yes, he was prepared to go along with these people for now, at least his reasons for tolerating Clell earlier still existed—he needed the help.

"It is difficult to trust someone I do not know, but I don't see as I have much choice," said Stan, smiling weakly.

"Have you guessed yet why we are here? You already know who we are."

"Yes, Crenums, it was something that Clell told me. Of course he also lied about being an agent for the UCP."

"Actually, Clell is a Zom operative, a very dangerous one with little regard for other people's lives. He has a well established reputation, and if you had killed him you would have done your world a huge service," he said as he shook his head. "I must admit that we were very worried for your safety when we discovered he was with you. Forgive me, I was about to say that you are lucky that we were here, but it rather looks like you were prepared to deal with him your own way, and our presence only served to complicate things. I am sorry about that."

"Don't be. I was equipped to kill him, just not all that sure if I wanted to."

"None of us expected someone like you to come out shooting—neither did Clell I think. It was a very dangerous gambit for you to point a gun at such a man as him, but it worked out almost the best way it could. Since we are trespassing here, as Clell said, and since our people are not actually at war with the Zom, there are some very specific rules we have to follow in how we handle him. None of us can legally harm him, and since he survived the shooting, we have to treat his injuries as best we can. To do otherwise would be…well…sticky; it would risk a war that I am under orders to try to prevent. We will take our time treating his injuries though, and then arrange for him to be returned to his people—after your ship is out of this place and we are all very far from here."

Grandin's face grew a little bit angry, "They go about, sometimes disguised to look like Telesians…Earthlings that is, sometimes not. Clell usually works in disguise too. His mission here was to alert his people, and yours, to the location of this place. If he had succeeded, all would have been lost."

"So they are working with my people?"

"Yes, with them...and against them. Your authorities do not know all that the Zom have been up too; if they did, I do not think they would be so cozy toward them. The Zom trample your laws as well as ours."

"What about this 'Mistaac Accord' of yours...aren't you breaking it too?"

Grandin smiled, "Well, I hope that with all of the evidence we have been collecting against the Zom, the UCP will be so outraged at *their* impudence, they will measure our activities here...lightly...on the balance. It is clear that the Zom have been exploiting certain exemptions in the law, exemptions that may now have to be amended out."

He slapped his thighs and stood up. "Right now though, we have your ship to finish. We need to get you and this vessel safely out of reach of Earth authorities, and quickly, and then out of Zom controlled space. If we can do that, then we might be able to get the UCP to intervene and prevent the Zom from doing any further damage to your culture. My men and I am here to provide you with the help you have lacked up to now."

He got out a pad of paper and a pen and handed it to Stan.

"I have to go to your house now, but first I need from you the locations of any research materials and personal items that you wish for us to secure. You are now in hiding, under our protection. I will not be able to allow you to return home again."

When Stan had written the information down, Grandin took the pad, put it in one of the pockets of his jumpsuit, and walked quickly toward the door.

"Get some rest. Our medic will help you fully recover from the effects of the stun. My men and I have already begun work, but we will need your help at some point." Then he stopped and came back over to

Stan, offering his hand to shake again, as if he was still getting used to the tradition.

"Team?"

Smiling, Stan grabbed his hand and shook it.

"Team."

***

Agent Mills stopped the CIA director in the hallway and handed him a printout.

Dear Gastone,

He is managed, and I am now safely in hiding.

Watch out for his relatives, they are not good people.

The CIA Director looked at this and frowned, then went into his office.

Chapter 5

"…Furthermore, it is the allegation and charge of the Crenum Alliance that the Zom Republic, did directly, specifically, and deliberately order the more advanced governments on Telis 3 to cancel all of their extra-orbital space flight research and development projects; and that they, and agents of those governments, have been clandestinely cooperating to suppress any new space-related research on Telis 3. We can only assume that the purpose of these outrageous interventions on the part of the Zom, was to undermine that planet's protected status under the Mistaac Accord…"

UCP General Advocate Steck Lertis held up her hand and Ambassador Klicktah stopped reading. She turned off the recording device on her desk and sat for a moment gazing at her copy of the document.

"These are very dangerous words, Ambassador," she said a little too crisply. "Is this a legal complaint, or a declaration of war?"

"What the Zom choose to do about these charges is up to them," Klicktah replied in clipped syllables. "But it is our opinion that their activities fragrantly violate both the letter and spirit of the Mistaac Accord, and we believe that most of the Council would agree. To answer your question though, it is not our position at this time to begin any kind of open hostilities with the Zom over this or any other issue. If the situation in the Telis system continues to deteriorate, however, the people of the Alliance will demand that our peaceful posture be reexamined. May I continue?"

"Just a moment, if you will," she leaned back in her chair. "This traditional reading the charges aloud is a necessary formality of this meeting, but it is only that. The reason we insist that you air them here first, in private, is to prevent embarrassment, and dangerous ignition of tempers, that can come from accusations that are either unfounded or impossible to prosecute. Are you sure that this is all just about the welfare of the people of Telis 3 and of the UCP, or might this have more to do with the proximity of the Telis star to Crenum space?"

"Well, I will not deny the strategic implications, but this complaint is only possible because of that 'proximity', as you so blithely put it. It has enabled us to collect the supporting data."

She thought for a minute.

"I understand. So you're claim is that you are bringing these charges purely out of the best interests of peace, and not for the advancing of your own influence in the region, even though you are using data collected through militarily funded espionage against a neighbor with whom you have, as you so blithely put it, a 'peaceful posture'."

Ambassador Klicktah was shocked. Despite the general advocate's reputation, he had expected more cooperation from her. He grunted and stiffly began to collect his papers as he stood to leave.

"Sit down Ambassador! You came into my office, and took up my time, and you will not leave until the meeting is over. Now I asked you a question, Sir. Answer it!"

He sat back down slowly.

"It is in the interests of all concerned, Madam Advocate, that these violations come to light. As for motivations, I can only speak for my own. The rest you know as well as I do; I assume that you have been following the press coverage."

"Ok," she said with mock politeness, "what are *your* motivations then?"

"I do not like to see primitive cultures so grievously tampered with. I also dislike seeing such uncivilized behavior and brazen disregard for the law."

"The law?" She stood and began pacing back and forth behind her desk, tapping her chin with her finger as she spoke. "Well, I believe you on the first point Ambassador, and for the record, I concur with you on the importance of preserving primitive cultures. As to the law, maybe you should know that the Zom delegate was in here just yesterday, sitting in that very same chair, complaining that your spies on Telis 3 are breaking the law, 'tampering' with the Telesians, undermining Zom interests and trampling on Zom sovereignty. He also sited the Mistaac Accord, just as you have—so forgive me if I'm not completely convinced of your sincerity with regard to a law abiding galaxy!"

Klicktah raised his brows.

"Oh don't worry!" Lertis waved her hand and sat back down. "His complaint is in my trash bin where it belongs; I laughed the arrogant twit out of my office. I wouldn't dream of putting my influence, not to mention the resources of my staff, behind such drivel. Most of his complaint was nothing more than an outrageous load of excrement, and I told him so. Oh, he didn't take it well, we shouted at each other for over an hour; you should have heard the awful language we used."

She leaned over her desk and grinned wickedly, whispering, as if to let Klicktah in on a secret.

"Incidentally, his un-researched allegations regarding my parentage were completely inaccurate—though my comments about his were probably not."

Klicktah was stunned. The way she whip-sawed…it was enough to give a person a neck injury. He didn't know what to think, much less what to say. He smiled uneasily.

"So…whose side are you actually on Madam Advocate?" he asked.

She leaned back in her chair, "Side? I am on the Council's side of course; they pay me to help keep the peace, and to prosecute violations of UCP laws and statutes. They *don't* pay me to prostitute this office, or it's resources, in the service of individual star nations. You've completely missed my point Ambassador; I am the first one who you must convince and I will decide whether your accusations are valid, substantiated, and if prosecuting them is in the best interests of the UCP. Only then will I prepare formal charges and present them before the Legal Committee. If I can convince them, then they will decide whether I take the matter before the Security Committee. Now, this matter would of course come before these Committees in closed session, at first, but the Zom delegate would be present."

She sat up, slapping her hands on her copy of the official complaint. "Your concerns over the Zom and Telis 3 are reasonable; I share them. I have been watching the situation there, and I am, personally, willing to believe these allegations…for now. However, the well-known strategic concerns, and the way you are willing to brush aside Crenum violations in this matter, bother me. The way I see it, these charges, combined with Crenum activities in Zom territory and all the posturing that's going on both sides, may only serve to provoke the Zom into taking preemptive action of some kind against either Telis 3, or the Crenum Alliance…or both."

Klicktah sighed.

"That possibility has not gone unnoticed by us, Madam Advocate."

"Good, then let me ask you some questions before we continue. I want you to know what you will be up against if and when we escalate this matter."

Klicktah nodded and waited while the General Advocate read the charges quietly to herself for a moment.

"I presume that you have good evidence of all this?" she said without looking up.

"We do."

"What kind?"

"Personal affidavits from our field agents, official Telesian documents, photographic images, recordings of intercepts…"

"Stop! Should I be hearing this? It sounds like a person would need a Crenum security clearance just to be discussing your evidence with you."

"Our documents, and my answers to you, have all been sanitized."

She shook her head.

"You see, that's a problem. In order for the Security Committee to consider your evidence credible, it has to be complete, in its original form, and include the identities and current whereabouts of individuals listed in the evidence, along with those who gathered and handled it. The committees will need to be able to summon and question them. That means that your government will have to release your field agents from their current duties and bring them here, which would expose them to capture and prosecution by the Zom for trespassing in their space. Don't think they won't do it! The Zom would immediately arrange to have them arrested and exterminated long before they would have a chance to testify."

"They would dare to murder witnesses? Crenum operatives?"

"Maybe not, but they would dare to 'accidentally' fly the prisoners into a star. Also, the original copies of these Telesian documents of yours would have to be examined, and certified by experts on Telesian culture and technology, most of who happen to be Zom."

"We have our own experts as well," Klicktah interjected.

"Right, but your experts and the Zom experts would cancel one and other out; and since the planet is in Zom space, and your experts gathered their knowledge while trespassing, they might end up being disqualified entirely, if they survive long enough for it to matter. I am not aware of any neutral authorities with experience in that culture, we will have to find some.

"There is also another issue with these documents," Lertis continued. "Any Telesian involved with them, as well as key persons shown in your photo images or heard in the recordings might have to be brought here to testify to the veracity of the evidence. Tell me, Ambassador, how do you propose to transport a group of Telesians here to be questioned before the committee? Are they protected by the Mistaac Accord or aren't they?"

"Advocate Lertis, of course we couldn't bring them here, you know that! You are being ridiculous!"

"Of course I am, Klicktah, after all I am not an ambassador—I am just a lawyer. It is my job to take the pathetic, superheated, mumbo-jumbo that you pompous, hotheaded, self-serving ambassadors give me, package it, and present it before still more pompous, hotheaded, self-serving ambassadors for debate and vote on. Politicians sat up there, made this law, and decided that it applied to any planet, without giving the slightest thought as to just how to detect violations and enforce the law inside a star nation's sovereign borders. Nobody gave a thought that someone like the Zom would look for loopholes! They expect this office to just fill in the blanks and take action! You are correct about one thing though, if it weren't for that planet's close proximity to your space or for your willingness to enter the system like you owned it..."

"We did use stealth ships..."

"Oh then, I apologize! You don't behave as if you own Telesia—you behave like pirates and smugglers instead! Please forgive me ambassador—I misspoke! Look, the issues that I am presenting to you here, in private, are eventually going to be thrust at me in public if this matter ever clears my desk. If we go to the Council with the full package of these charges, it will embarrass, antagonize, and infuriate the Zom well beyond their very limited capacity to be reasonable, and frankly, I fear for you and your people. The Zom would feel perfectly comfortable using your trespassing as a pretext to start shooting at Crenum ships, just to muddy the waters, and I have no power to help you. I command no fleets to send to your aid. I can only present the case, such as it is, before the star nations represented on the Council. They wield the real power, but they need something to light a fire under them and their people or they will do nothing. You and I both know that the Zom and their friends have the military capability to wipe up the walls with you and unless your complaint wins support, you will receive no help from anyone. There are really only two factions to this issue, and right now, those who agree with the Zom may be the only ones with any backbone. You see. That is why your evidence must be presentable. Now, do you have copies of this evidence with you?"

"Yes, I told you that when I first came in."

"Of course, now let's sit down and go over it item by item; the formal reading of the charges can wait. If any of this turns out to be credible, then those cowards on the Council might finally acknowledge the Zom's' continued, reprehensible disregard for civilized behavior. It constitutes a direct challenge to the authority of this body and its capability to enforce its laws, but none of that will ever matter unless we can get them to take quick action to stop whatever atrocities the Zom are planning to do next. You see Ambassador; I happen to know about that ship your people are fiddling with on Telis 3, perhaps even more than you do. Trust me, that whole project hangs by a thread, and most of this case hangs with it. If your people can't get that ship out of there, little else will be possible. If the Zom, or the Telesians catch them…"

What?

"Madam Advocate, I can't discuss the specifics of any of the activities that our operatives may or may not be engaged in," Klicktah put in.

"Delude yourself all you want Ambassador, but I need you to document where and how you collected all of this evidence, who collected it, and where they are now," she said, straightening some things on her desk. "Eventually, I will also need written details on this McPherson spacecraft, and just how much your people have tampered with it."

How did she know so much!

She drew back and exhaled and a fire lit in her eyes behind welling tears. "I know what the Zom are doing to Telesia," she said with a sneer, her voice shaking, "and I hate it; they and their treatment of primitive worlds disgust me. If this situation is salvageable, the charges substantiated, the evidence genuine, and the case prosecutable, I promise you that I will risk my career and move the stars themselves to serve the law and stop them. I want to protect that planet too, but believe me when I tell you that my deepest duty to the UCP lies with the preservation of peace. I cannot have Zom evil spreading outside their borders. How many developing worlds do the Crenum protect, Ambassador? Would you like to see the

Zom gobble up those planets too…enslave them the same way they plan to enslave Telesia? The Crenum Alliance is very important to protecting developing cultures and maintaining peace in that region. If the Zom destroy you, who will be left to contain them? If I think that this chain of events will not save Telis 3 but will only end in the violent overthrow of the Crenum Alliance by the Zom, then I will end all of this! This issue will die on my desk! I am as frustrated with the Zom as you are, but to preserve the peace and protect the good people of the Crenum Alliance, and others, I would sooner see Telis 3 on fire, your complaint in the trash, and your spies in chains!"

For the first time in weeks, Klicktah smiled. His voice was calm.

"Madam Advocate, I came here expecting nothing less."

***

"If this ship is going to serve as proof of Earth technology, won't the work you folks do to it make it seem more like you built it instead of me?"

Stan looked over the shoulder of one of the alien technicians for a moment as she installed some equipment. It amazed him; in just a day of asking and answering questions, he'd learned more about his propulsion system than he could have researched on his own in a year.

Grandin took him gently by the shoulder and turned him away. They continued their walk.

"No," he said, "we are keeping a record of all of the work we do. We intend to follow your original designs, and have planned few enhancements. We will use composite pre-forms from your own stockpiles whenever possible and we will use your sources, and low-tech Crenum worlds, for anything you don't already have.

"We hold encroachment meetings anytime we feel a need to use any of our material, equipment, or technology, and document any variance from your plan. However, you have already completed the drive ring without our help; that alone is the biggest and most relevant proof of your planet's status under our laws. The most critical task is to move that evidence of your achievements to a place of safety," he grinned. "We think it's worth a little cheating."

***

As the days progressed, the tempo of the work quickened. Every morning Col. Grandin met with Stan to report progress and discuss general issues, just before all of the Crenum got together in a separate meeting behind closed doors, discussing things related to their own science, security, and technology.

Work that Stan had planned to take months to complete, took weeks or days. Stan would walk by and see a group of aliens working on something that he had looked forward to completing and when he returned the next day, he would find the task completed. The aliens sometimes brought equipment, some materials, and other supplies in and out of the facility aboard several of their own spacecraft—which they did not permit Stan to see. Construction materials that they shipped-in arrived pre-milled to the desired shape, and were indistinguishable from materials he prepared himself, except for their quality. Working alone, and in secret, he had to tolerate lower quality composites, because they were all he could obtain quietly. The materials that the aliens brought matched the standards that NASA contractors would have followed—if NASA still existed.

They needed Stan to help them on certain details at first, but eventually they learned his design through their interviews with him and by studying his ship and notes. Stan soon found himself with little to do except wander the facility and ship, watching them work.

As the work progressed, they held their encroachment meetings with greater frequency. Stan could hear them shouting at each other as they debated.

Everywhere Stan went, he saw Crenum technicians and equipment. Sometimes there were empty spaces filled with strange devices, or holes cut in walls and alien technology installed there. Sometimes, he would find work that he had sweated over with much care and tears, torn apart and rebuilt in such a way that he no longer recognized or understood it.

He could not even send out his probes anymore—doing so just seemed unnecessary and dangerous under the circumstances. Sequestered in the facility, he no longer had the ability to retrieve them from remote locations, as he had done in the past, and programming them to return directly to him was unthinkable.

The aliens were very polite, greeting him warmly and speaking English to one another whenever he was around. They all seemed to have a strong regard for him, revering him and treating him as if he were some great inventor from their own history, despite the fact that, compared to them, he was a mere child in school. Sometimes he could get them to loosen up a little and tell him about places in space where they had been, things they had done, or glean little snippets of their technology. They were careful though, sharing some science with him while kindly and tactfully refusing to tell him other things. They seemed to have "talking points" that they were specifically permitted to discuss. The scientist and astronomer in him found their tight-lipped posture maddening.

He began to feel like a guest on his own project. Up until then it had been just him and the ship, as he carefully fashioned every part with the love of a parent. The project had been the main reason for his existence for more than a decade, even before the government disbanded NASA, but as the Crenum workers took over he stopped feeling a part of it. Every corner had contained a small part of himself—now these smiling, polite aliens were slowly smudging out all of his fingerprints. He spent his days wandering, trying to be helpful and trying, and failing, to keep up with the progress. The ship, to him, had stopped being his. It also started to look more alien than Earthling, at least on the inside.

He explained his dilemma to Col. Grandin.

"Well, there are places where you followed errors in concept that we had to correct in order to make your work compatible with ours. Don't feel bad, our worlds all made the same mistakes when we first started."

"I'm sure they did, but they probably got to learn, improve, and correct such things on their own, as they went."

"Well, usually yes. They also had entire industries helping them, experts to innovate solutions, and consumers to test new things and point out the flaws. There were also spin off discoveries that enabled further improvements. You know how that works."

"Ya, like the development of microcomputers at the end of the 20th century here on Earth."

"Exactly, I read about that. It was the same with your invention of the automobile, and the airplane. Each discovery influences innovations in other fields, which in turn empower more discoveries, each invention stitching itself into the fabric of your culture. This is how all technologies develop on your world and on mine." He shook his head. "You don't have that kind of help from your people. I have been trying to work around it, but sometimes we have to simulate it. We have to ask, and answer, all of the questions the way we think your world would have answered them. Space flight will never be for you the way it was for us, and if the Zom invade, it will be far worse. That is why our most critical objective is to keep the drive ring authentic and get it off this planet. Someone even suggested that we just scrap the ship and haul the drive ring off in one of our own freighters, but I would not allow that. This ship belongs to your world, and it would be wrong for us to destroy it. We would be no better than the Zom if we did that. However, we are short on time, and it has caused us to have to take many shortcuts, too many I think. I am sorry, but Zom interference with your world has stolen a lot from your people."

"There are no Zom with us here in this cave," Stan said.

"That is not what I mean; I am referring to how they have forced your people to cancel the science programs that would have allowed your work to progress normally. NASA should have completed this prototype long ago. Your people should have evolved this science properly and enjoyed all of its spin-off benefits—if it weren't for the Zom treaty…"

"What treaty?"

Grandin stopped, and then shook his head.

"I am sorry, I thought you had learned of that by now."

"Well, I've figured out that they have some kind of contact within my government and that agents like Clell suppress innovation by killing scientists and others who get in the way."

"Connect it together, Stan," said Grandin, his voice grim. "Who makes and enforces the tax code? The Zom don't do that. Who buys up and buries pro-space flight patents that get past the Zom? Who guides the path of your science fiction literature away from space science projections? Who popularized this new, negative view of the space oriented science fiction of the past?"

"Well," Stan pondered this. "The people and the government did those things."

"It would be an interesting coincidence if they did," Grandin shook his head. "Your people turned away from space flight at the same time that the Zom secretly arrived and started interfering. The truth is that you're people would *never* have done this to themselves otherwise."

"Your opinion."

"It is the opinion of a broad group of my peoples' cultural scientists. Look, Stan, the dominant culture on this planet still consists of people like you, innovators...explorers. Telesians are still very technologically aggressive. Such people rarely become permanently grounded, and your people probably would not be grounded now, without some kind of focused plan to keep them disinterested in space."

"A government conspiracy?"

"Yes."

"That's not possible."

"I've seen the documents...and the signatures."

The way Stan remembered events, the people had risen up and demanded that science look more Earthward.

"You mean the Zom and the U.S. Government *agreed* to work together, to suppress interest in space?"

"I guess I have said too much."

"No you haven't. I did hear rumors, years ago, about something like that, but I didn't believe it. There's too much transparency in our system for it to exist for so long. As for alien collaboration, that would be explosive, and very difficult to contain."

Grandin thought for a minute.

"The treaty I spoke of comes in the form of a peace treaty your government signed with the Zom fifteen years ago...a treaty which forbids your people from pursuing any space related technologies which extend outside close orbit with your planet. The Zom threatened to invade if your government did not sign this treaty. They twisted the meaning of the Mistaac Accord and said that it was illegal for a culture in your stage of development to have space technologies. They told them… that in order to remain sovereign, they would have to abandon those sciences for a while. It is like the nuclear non-proliferation treaties that you still have with some of your less developed countries."

"Such things tend to leak."

"The Zom have ways to manage that as well…"

Stan felt a chill. He thought about the day that Clell had showed up by the pool with a briefcase in his hand, offering to help him.

"Clell."

"That is what Clell does. If it were not for the treaty, NASA would have built this ship, and other ships like it would have made mining and commerce on your moon and asteroids a routine. There would have been a Helium 3 isotope mining industry born to fuel them. At that point the Mistaac Accord would have shackled the Zom with respect to you and placed your world under the direct oversight of the UCP. Your industrial age, and your space age, would have invigorated one and other. Fueled by helium 3 nuclear fusion they would have propelled a new age of interstellar exploration, eventually bringing you into contact with alien races. You would probably have joined the Crenum Alliance instead of the Zom Republic, since our policies place a higher emphasis on allowing worlds to govern themselves.

"That is why they forced the treaty, and why they enforce it. They are under some pressure diplomatically right now with regard to Telis 3—the Crenum are applying that pressure. The Zom found out about your ship, but had no idea how far it was from completion, or where it was. If you had finished and flown it, the Zom could not have continued to claim that your world has grounded itself. Clell contacted you to prevent that."

"As it is, his mission hasn't necessarily failed," Stan said thoughtfully.

"No, it has. We have him under deep sedation and we can keep him like that for a very long time without..."

"That's not what I mean. From the perspective of preventing this world from advancing on its own, his people's interference has forced you to interfere."

"That is true," Grandin acknowledged.

"So one form of interference has been replaced by another."

"What?"

"Well, you and your people watched me build my ship for five years, without introducing yourselves or helping. That changed when it looked like the project would be lost."

"True, but we are helping, not interfering."

"You are helping to build the ship, but you have to interfere to do it."

"It is necessary, the Zom have seen to that."

"I know. We are where we are, you and I, because of the Zom. The way things are going though, I am no longer building this ship—you are. It is no longer my ship, but yours. Don't you see? From a certain perspective, Crenum agents are on Telesian soil, building a ship out of Telesian materials—and simulated Telesian materials—patting yourselves on the back and calling it 'Telesian' and 'proof of Telesian advancement'. You have completely pushed me out of the process. When we leave here, it will be to deliver the ship to the governments of those that built it. The one Telesian who could be involved has been shunted aside. It will not save Earth, because it will no longer be Telesian, but Crenum and the last piece of Telesian space technology will be gone. Those who support the Zom will declare this project a fraud."

"We cannot involve you further than we have, we are walking the fine line of the law as it is. There are things that we simply cannot tell you and other things that we have not time to teach you."

"I must be able to understand and maintain this ship's systems when all of this is over. I will need to be familiar with the full breadth of your technology, at least so far as it relates to equipment aboard this ship. I understand the reasons why you are doing things the way you are, and I agree with your reasoning, and I appreciate the help, but the staggering pace at which you have accelerated past my work has me bulldozed."

Grandin sighed. He took a document off his desk and showed it to Stan. As he read it, he tensed; it was an intercepted C.I.A. document discussing his business and real estate holdings. Grandin took the document back and set it on his desk.

"As you can see, we don't have a lot of time. The Zom are pressuring your people to find and eliminate this project, saying that it is a violation of the anti-spaceflight treaty. I am personally amazed that your people have not found us already. My people have begun constructing a hidden facility in a friendlier country, in case we have to leave here before the ship is space-worthy. We clearly have no time to keep you caught up on all of the things that we are doing, nor can we hold your hand through what should have been many years of technology advances. Relax!" He put his hand on Stan's shoulder. "We all want the same things."

Then why did he feel so alone?

The ship had been his life's work. He had no family, saw few of his old friends, and pursued no other hobbies. He had grown accustomed to being alone, working alone. He felt like one of those people who had been born and raised in one of those slow, small towns and then moved away to the city, to find masses of people, rushing around and doing things that he didn't understand, at an urgency and pace that he couldn't comprehend. He found himself wanting his solitude back.

Throughout his career, he had been a socialite, and as an astronaut he had been a public figure; recognized, it seemed, by all the world, and liking it. Then that same world had forced him to become a hermit, alone with his ship. Now he had companions, but he was still alone.

Then he realized something he hadn't considered before. He had looked forward to the solitude of space flight as well. He had grown so accustomed to doing everything himself, that he'd grown to like it too much.

The flight into space would have been a different story. If Stan had attempted to fly alone into deep space, he would have gone mad. He had been looking forward to the flight as a solo, but if he had tried it, it would have been disastrous. He had flown in space before, yet all, even in the close Earth orbit that the Space Shuttle and the International Space Station had used, had felt the loneliness of space. When he had flown to the Mars "Space Loneliness" had been a serious study and acquired several clinical names during the planning for that mission. All life forms have evolved in them a "feeling" for connection with other life, so that when anyone was in space for long periods, looking out a view port into the darkness, he or she often felt a loneliness within that was as vast as the emptiness without.

NASA planned to solve this problem by putting crews together months before a flight, to live together, eat together, and work together. Those who did not "fit-in" to the group would be switched out. Immense databases of Earth cultural materials were to be brought along for diversion, to remind people of home. NASA even toyed with the idea of permitting pets on board, and plants, so that no one would feel a lack of life around them.

"I completely understand Colonel, but this is still an Earth ship. There is only one Earthling here with you, and that Earthling feels like he is just in the way. You have placed my name on the list of crew members, yet I am not actually a part of your crew."

"Mr. McPherson," Grandin said gruffly, "Haven't you been listening? We have days instead of weeks, and minutes instead of hours to get this ship out of here. The danger to you and to this world is too great to delay. Once we are in space and on our way, there will be plenty of time to catch you up, believe me."

"You talk as if you have done this sort of thing before."

"Nothing remotely like this has ever been done before."

"Then how can you presume that you know what is best. I'm sure you are familiar with space loneliness?"

Grandin stopped in mid-breath. He seemed to think about that point for a while, and then he sighed.

"My father was alone in a life pod for four months. He told me about it. We have medical disciplines for dealing with what we call 'Life Lack Depression'".

"So do we."

There was another long pause while Grandin thought this over.

"I will get back to you on this."

Grandin refused to discuss the matter further that day, but the next morning Stan was invited to participate in the morning planning meeting, which was held in English for his sake. The project managers assigned him study and work duties and expected him, even ordered him, to complete those assignments. He immersed himself in the difficult work with vigor, not wanting to disappoint them. Their expectations were as high as their regard for him, and he completed his assignments on time. He wanted to earn the respect that they held for him so generously.

Stan could tell that they sometimes sought, and were granted, permission to discuss sensitive technology and other issues with him. One day, Stan asked someone how their equipment was to be powered, whether by the power systems that he had built, or by the alien power units for which their equipment was designed. The technician immediately looked uncomfortable and politely changed the subject. Several days later, this same alien took him aside and handed him a diagram. He then brought Stan to the cargo bay and showed him a piece of equipment that the Cre-num had assembled.

"You can see this device in the plans there. All of the non-Telesian equipment that we have installed, are designed to be add-ons to any small or medium sized vessel. We have brought and installed a reactor, which is quite small and portable by your standards, that powers all of our equipment. We installed it elsewhere in the facility for now; but we will move it aboard the ship before flight. We provide the fuel that it needs, but in flight we will need to be able to manufacture more of our fuel from yours. This piece of equipment is a fuel converter that will take your cryogenic Helium 3 isotope and adapt it to the needs of our reactor, which uses a chemically packaged form of H3 that can be stored at room temperature."

He pointed on the drawing, moving his finger in the direction of the flow of the fuel.

"As you can see here, this converter will be installed near your main storage tank and will take your cryogenic H3, convert it to our fuel, and then pump it through a tube into the cargo bay, where it will be stored again in a separate tank, and used by our reactor."

Stan looked up from the drawing and glanced at the spot in the cargo bay where the alien reactor was to be installed.

The alien was right; it was very small, less than a tenth the size of the two redundant reactors that he had. He admired the way alien technology solved problems he had puzzled about, sometimes for years. Storing and managing cryogenic fuel was a dangerous headache and the equipment necessary to handle it and keep it at such cold temperatures was a waste of power and space.

I wonder how they do it?

Chapter 6

"I have a surprise for you!" Grandin said, interrupting Stan's studies. "Come, please."

They walked to another area of the facility, a warehouse that Stan had used to receive and store construction materials. Stan was pleasantly shocked to see an alien transport ship parked there; they had never permitted him to view their ships. It was wedge shaped, something like a thin slice of a thick pie, except that it had a narrow, PAGA drive ring encircling the back and sides of it. Stan was struck with two things, first that the ring did not form a complete circle around the ship—the nose of the vessel protruded from the front portion of it—and also that the ring was very small, about a forth the mass that he would have thought was needed for a ship that size. He pondered these issues for a moment.

"Nice ship," he said, "yours?"

Grandin regarded him closely for a moment before responding.

"The ship is not the surprise" he said, somewhat gruffly, "and I am not going to answer any questions about its design, so don't ask." Then he looked back toward the transport and smiled a little.

"That drive ring seems kind of small."

Grandin looked back over at him with his eyebrows up and said with emphasis, "*Inside* the transport are some Earth weapons that we have stolen from your country's navy, to mount on your ship."

"Earth weaponry?"

"Yes, it is necessary that the ship go armed, since it may come into conflict with either Zom or Earth forces when we try to move it. However, we cannot equip you with Crenum weapons, because doing so would be illegal. We cannot legally even bring such things within your planet's orbit. This equipment that we have 'acquired' is some of your country's latest and most sophisticated laser and missile weaponry and counter measures."

Stan thought about that for a moment. "Earth technology they may be, but I know very little about the newer weapons that my country uses. I used to be a fighter pilot, but that was a long time ago. Who will install this equipment?"

"I will," said a woman's voice behind him.

Stan turned, surprised, and saw a lovely human woman, in her early 30s, standing next to him. She wore a close fitting Crenum military jumpsuit with a Crenum pistol in a modified Army-issue holster slung around her waist.

Grandin turned and gestured to the new arrival.

"Allow me to introduce, former military defense contractor…"

"Merry Christmas, Astronaut Stan McPherson! I'm Dr. Claire Saunders," she said, grabbing his hand and shaking it in both of hers, smiling. "I have been looking forward to meeting you in person; I've heard and read so much about you." She took him by the elbow and started dragging toward the alien ship.

"I went to some trouble to get you these toys, McPherson," she said with a mischievous smile. "Sorry there wasn't time to have them gift wrapped."

So that's the surprise, thought Stan, looking back at Grandin as Claire hustled him aboard.

***

"We've put together a pair of rotating weapons rings to fit above and below the drive ring," she explained. "That'll give the ship a full field of fire. We'll put the Phalanx laser clusters here, here, and here on both rings."

After going over the weapons inventory, Stan and Claire ate a late breakfast in the old break room of the facility. It was a large room, designed for a hundred employees to take their breaks and lunches, and had more recently served as a very quiet, lonesome place for one man to eat. Claire was still going over the plans, drawing them on the back of a paper plate.

Claire had worked with the Crenum much longer than Stan had, so he knew there were many questions he should ask her, but he couldn't think of what they were just yet. He was still getting over the shock of encountering a fellow Earthling in league with the aliens. It left him off balance. He began to wonder if her good looks might not just be some kind of demented illusion on his part—a product of his shaken psyche. He was sure that Grandin and his people had sprung that part of the surprise on him on purpose.

"How long do you think everything'll take?"

"I'll start running the computer links right after breakfast," she said. "It shouldn't take us more than a few days to get everything set up—I already updated the ship's combat computer last night. The targeting software that I brought is the latest and greatest. Oh, and I had a chance to play with your simulator last night too…sweet!"

Stan grinned, "I'm glad you like it, it was produced by an offshoot of your company…or the company you used to work for that is."

"Ya, small world." A ghost of sadness flickered across her eyes, and then passed. "Listen, Stan, I'm a little bit out of touch with developments lately...been busy. What's news?"

"Well, the ship is due to be repossessed by the IRS, the NSA, the CIA, or the Zom, any day now. We leave lights off and the curtains pulled—so no one can find us home to serve us papers."

She laughed. He wondered why the sound felt so pleasant in his ears.

"No, seriously, outer construction is nearly complete, as you probably saw. We haven't yet finished installing and testing all the avionics, but the ship will be flight-capable very soon. We can't have an actual test flight until we're ready to go for real though. If anyone sees us lift off we can't come back."

"Oh."

"I never dreamed of getting this far so fast, having these helpers has been a miracle...it's just too bad we can't show it off."

"It wouldn't wake people up, Stan," Claire looked at him, "even if you flew it right over the Super Bowl at half time. They've made their decision.

"I think they'd come around, with time."

"They're out of time."

"They just need some outside help—like I did."

"Ya," she looked away, "the Crenum do that a lot."

Stan looked at the eggs on his plate. He hated to steer the conversation to the negative this way, but confiding his innermost fears to aliens was still hard. He needed to bounce some thoughts off her—hear how they sounded coming out of his mouth.

"If the Zom come, the world is toast," he said into the silence. "Even though there's still a lot of technology these Crenum hide from me, the little bit I do see is more than I can comprehend. They're installing their type of shields on the ship, and I have no idea how they work, but from what I've seen demonstrated I can't see any way conventional Earth weapons can penetrate it."

"Well, we no longer have just lasers and chemical explosives, you know," said Clair. "The U.S., and some of their allies, and China, are just now coming out with some particle beam and plasma technology. That stuff can damage alien shields, though not very quickly. Either way, defending against alien weapons without shields is almost impossible. At close quarters, most modern fighter aircraft could maneuver fairly well against alien fighters but their range and top speed can't touch anything with a PAGA drive and shaped shields. I've seen more of their technology than any other Telesian—much more than Grandin would like me too, and I agree with you; if the Zom come here in force I don't know how Earth could possibly survive."

"Nukes?"

She shook her head, "We can't use regular nukes even if we wanted to—we still have to deliver them to the target the old fashioned way, and then the alien shields are designed to deflect most of it. We would have to get the warheads close enough for numerous, precise hits, but we've built all our stuff to attack Earth targets, not big, mean things coming in from the outside. At least we're in good company though; Grandin tells me that people like us on other worlds tend to build that way too…prepare only to kill each other."

"How did you get involved with Grandin anyway?"

"The Crenum saved my life." She chuckled a little. "My team had a space-based weapon proposal squashed. I thought maybe a rival company had greased some palms or something, so I looked into it and stumbled across that secret deal our government made with the Zom. I tried to expose everything to the press and found out just how unhealthy that can be."

"Ya, people die from that," Stan said with a sigh. Memories of lost friends came flooding back.

"Yep, so I got scared and went into hiding. Everybody was looking for me, CIA, FBI, but I hide well. No one could find me. Grandin and his people were searching too, they say they couldn't have me going public either, and they wanted to get to the evidence before the Zom did. That and they don't like that little fetish the Zom seem to have for making people disappear."

Stan sipped his coffee. He had only recently come to fully believe in the conspiracy, and was still, slowly, working through his memory of events—examining them through the new lens that involved an alien tyranny, and people like Clell.

"Why would the Crenum care if you went public?"

"Well, as it turns out, it would be a disaster, because if people believed me, then the Zom reaction would be to immediately invade, to keep the situation from unraveling further and not give Earth time to prepare to fight back. They have stuck their fingers so deep into our government and industry, they could mess us up from the inside out while their space fleets came in and took over. Grandin says that's been their pattern on other worlds. It makes sense really; what they want from our planet wouldn't work as well after a messy fight.

"Anyway, Clell found me first. He lured me out of hiding, claiming to be a rival alien to the Zom, even offered to take me off-planet somewhere. Like a dork, I believed him. Men! When it finally dawned on me what Clell was, we had a little fight. I turned out to be harder to hold on to than he expected and he turned out to be better armed than I expected. He hit my truck with some alien EMP thing—froze the engine dead! It happened while he was chasing me over a mountain pass in Tennessee and I went off a cliff and rolled. When I woke up I was on a Crenum shuttle."

"How long ago was that?"

"About three years, I've been working for Grandin ever since. I help them hide from our people, and teach them English and American history and culture and stuff. I also do intel for them, and help them get things they need. In exchange, they treat me pretty much like one of them. I speak their language now, and I research their technology and culture every chance I get. That's what I do best, research and acquisition." She looked at him and tossed her head with mock arrogance. "I've made myself Telesia's leading human expert on alien toys, but this past couple of months I've been busy getting Earth weapons for you."

"What about your other life?"

"What other life? I can't go back, Stan. Working for Grandin is the only way I can fight the Zom and continue walking around free and breathing. I *have* to fight the Zom now, my old life, all the things that I used to do, just aren't important to me anymore. I can't just go on building and selling weapons for Telesians to kill other Telesians, not when I know the Zom are getting ready to blow us all to Hell. It just wouldn't feel right. I've dumped everything else in my life that used to matter, including my patriotism. I no longer consider myself loyal to the United States, I'm too anti-Zom to be an American anymore. For me there is no other issue, it is my own little personal Crusade. I call myself a Telesian, not 'Earthling', but lately I think I'm more Crenum than Telesian. I live in hiding, I can't come out, and the Crenum wouldn't want me to anyway—I'm useful to them." She looked at her plate. "And I know too much."

Technically, Stan had allied himself with a foreign power too, one who opposed current government policy. Since the United States had been cooperating with the Zom, they might take the view that the Crenum were an enemy force. He had shot a Zom representative. Did that make him a traitor? Events had guided him more than he had guided events. Besides, would not his actions, like the actions of these aliens who were with him, be justified in light of Zom crimes and future intentions? Did his activities make him less American? To his mind, he was fighting for his country and his way of life.

"Claire, why do you fight the Zom?"

"Well, because I resent what they're doing, silly."

"So it isn't just because they tried to kill you?"

"Of course not."

"Well then, you shouldn't consider yourself anti-American. You're a patriot. You're trying to liberate your country."

A sly grin spread across her face.

"Is that before or after I 'liberated' all that military hardware back there and loaded it aboard an alien ship to give to an outlaw?"

Stan grinned at that.

"Speaking of outlaws, Claire, just where did you happen to come by all that fancy weaponry?"

"That's my little secret and I'm not going to share it with you," she said, sticking out her tongue. "But I can get almost anything—I have connections. I know when and where to find things, and I have lots of alien gadgets to help me snatch what I want and scurry away with it. Radar and graviton sensors can't pick up that cool stealth freighter they let me use, and I get to bring along lots of big, strong, alien men to load and unload my plunder for me. I'm nearly invincible."

"They told me about those stealth shields," said Stan. "They think they've found a way to bend that technology to fit around my huge, dinosaur of a ship."

"It's never been done before, *ever*, not with a ship the size of yours. The Crenum are the only ones who know how to do that to a ship larger than a fighter."

She wrinkled her nose at him and punched him on the shoulder, "Hey! My food's getting cold; you talk now while I eat. What is this place, anyway? It's like working for Batman."

"It belonged to one of my dad's companies; he was somewhat of a visionary. He saw the potential of Helium 3 fusion energy, and this used to be an H3 research facility. Trouble is, there's not enough H3 on the planet to support an industry. The best, closest source for H3 is the moon." He pointed upward. "When the space program went under, the future of H3 power went with it."

"Ya, well…no one said we Telesians had much in the way of brains."

"It's all so frustrating. I started building this ship to help the world; I guess saving it from an alien invasion is just as good. It's great to have some support though, it was turning out to be a daunting, and very lonely task. It was also far more expensive than I thought it'd be."

"Ya, I hear you're rich, you can afford it."

"Well yes, I am rich. I have a big house, lots of land, a big swimming pool—"

"Awesome, let's go for a swim!"

He smiled.

"Well sure, if you want the Zom, the CIA, the Crenum and who knows what all else to see what that thirty-something body of yours looks like in a swimsuit."

She thought for a moment, and then wrinkled her nose again. "Well, I'd better pass on the swim then…" she patted her stomach, "…still carrying some Christmas flab."

***

About a week later, Grandin called an emergency afternoon meeting.

"I have just received word that the evidence we collected against the Zom was used to file formal charges, along with a cease and desist order, with the UCP Legal Committee. They did this over my recommendations that they postpone such action until we complete our work here. Now, since the Zom are unable to get any traction in the UCP with *their* complaints against *us*, they have decided to react militarily. Word has it that the Zom have dispatched their 3rd and 8th Fleets in haste from the Regitus system for parts unknown, presumably for points nearer to the Crenum border.

"The Crenum Governing Council has called on the UCP to put this area directly under their control, saying that 'the Zom seem incapable of managing Telesia properly', and have ordered the Crenum Defense Director to prepare a plan to directly intervene here in the event the UCP turns out to be unwilling, or unable, to enforce it's laws. However, it seems unlikely that our government is serious about actually deploying forces here. Most back home are taking a more passive stance, afraid to antagonize the Zom further. The Zom have declared that they will 'resist any further encroachments of their space with unrestrained force' and have hinted in some of their rhetoric at the possibility of a direct attack against the Crenum Alliance itself. Our own 1st fleet, and other units, are being mustered to support our more vulnerable outposts, so most of our newer vessels will be spread out, to strengthen systems near the border." He shook his head. "Fortunately, at least for the moment, the Zom still seem to be keeping their available strength in this system down to Mistaac Accord levels, but they are now patrolling the large outer planets more aggressively, to try and catch our transports moving in or out of the system.

"The U.S. government has the ability to find us very quickly, since this location is part of McPherson's very well documented financial property, but they don't seem to care as much about us as the Zom do. Our sources say that the Zom know we are here, they know what we are doing, and have redoubled their pressure on the U.S. Government to catch us and terminate this project as soon as possible. If they find us, the Zom will destroy this ship and distort any proof of our presence here to leverage a favorable result in the UCP courts and justify them in invading the planet, in the name of the Mistaac Accord, to 'cleanse' it of alien influences. They know that to prevent that, we would have to evacuate in advance of any Telesian operation against this base and leave McPherson here with his ship to fend for himself."

"You said that they know we're here?" Stan asked.

"Forgive me, let me clarify. The Zom know about this operation. They know that there is a sizable Crenum reconnaissance force operating from various hidden locations within the system. They also have pretty much figured out that we are helping you finish this ship in an effort to move it out of their reach. They are scouring both the planet and the system for any sign of Crenum ships or bases, and for this base in particular.

"We are too vulnerable here. My new orders are to get this operation mobile as soon as possible. We have to bring this ship up to atmospheric flight worthiness immediately and leave this place at once. All work unrelated to that end is to be stabilized, secured in the cargo bay, and resumed at the new location. Be careful not to leave anything behind that could indicate that outworlders have been here, no matter how small. There are rodent scavengers that live in places like this that like to carry off small things and hide them, so scan everywhere. We are in a hurry, but we cannot afford to be careless. As soon as all of the systems that you oversee are ready, move them aboard and turn them over to the cargo master. All sections report your flight readiness to me by tomorrow morning; we fly tomorrow at sunset. All personnel, goods, and equipment that will not be going out on this ship need to be made ready to leave on the transport at midnight tonight, or later on the shuttle. Claire will be running a little errand in the shuttle tonight but it should be back in time to depart here at around 4:00 am tomorrow morning."

As the meeting dispersed, Stan pulled Col. Grandin aside.

"Where is it we're going?"

"My people have established a base elsewhere on the planet; they are preparing at this moment to receive us. They tell me that it is well hidden, pretty much out of reach of your government and very difficult for the Zom to find."

"Where is it?"

Grandin hesitated a moment, then looked at Stan carefully,

"I am sorry, but they are still containing that information; even I do not know exactly. You will be piloting the craft, so you will know soon enough. For one of us to fly your ship under these circumstances would be inappropriate. It is complicated, but if any of your people were to see this ship while you are piloting it, it is a different matter than if one of us is the pilot. In fact, since you and your ship are of this world, I am going to put you in full command of this operation from take-off to landing. That way we are not stealing the ship away, but rather you are fleeing to us. I know it sounds silly, but it is an important distinction."

"What if we're attacked?"

"Respond as you like," Grandin said, shrugging his shoulders, "You will be flying the ship while Claire functions as scanner and weapons officer. You designed it so you could operate it with a very small crew anyway. The rest of us will remain aboard to monitor the more troublesome systems, make repairs, and solve technical issues. I can advise you on tactics, but you are a combat pilot and Claire is an expert on Telesian weapons. When it comes to flying and fighting that vessel, Claire and yourself are the most qualified team available to do that. Besides, we Crenum cannot do any harm to your people or the Zom, not even in self-defense, but you and Claire can. The Zom are definitely your enemy, even if your country does not recognize them as such, so if you fire on a Zom ship, then it is not that much different from when you shot Clell. If a Crenum commanded vessel were to fire on a Zom ship it could trigger a war, and more than likely spark the very invasion of Earth that we are trying to prevent. We can support you and would enjoy watching the battle, but we cannot pull the trigger. As for any Telesian forces that try to interfere with us, they are your people, deal with them in whatever way you deem appropriate. Just know that if they capture the ship, with us aboard, they will hand us over to the Zom, and the Zom will twist the purposes of our presence here in such a way so as to justify them in taking full control of the planet."

"Of course."

"All the same, I do advise that any weapons fire be used with restraint. That ship is not a warship, and is much better at running and hiding than it is at fighting. It should be very hard to detect for your people—and even for the Zom—because of the stealth equipment we installed, but your weapons probably wouldn't do any damage to a Zom ship anyway. The Mistaac Accord will not keep them from following us, and they might even attack us, if they think they can do it quietly. Either way, we cannot allow a Zom ship to follow us to our destination, so we will have to have a plan or two in place for dealing with them either way.

"As for Telesian forces, if any of your own country's military aircraft were to swarm us, they could make it impossible to get anywhere. Yours is a civilized nation, so they will probably not attack you outright, but if you fire on them I think they would react very violently and some of your planet's fighter aircraft have weapons that can damage our shields."

"I will not fire on any U.S. Forces anyway, not even in self defense, but I will attack drones or missiles if I have to," Stan chilled at the thought of killing those brave young soldiers.

"I cannot advise you further. Please write up your rules of engagement and bring them to the morning planning meeting tomorrow, so that I can brief everyone on what to expect."

Chapter 7

Later, Stan and Claire got together over dinner to discuss tactics.

"Ok, Claire, I haven't flown in a fighter since 2008. I've heard that the military has done a lot with artificial gravity; you know the details better than I."

"Well, you know I ripped out those cheap, after-market high-G compensators you installed and replaced them with the best our country makes. The Crenum ones are much better even than that, but their equipment is going out on the transport tonight; there was no time to install it. We have the PAGA drive, so if Earth fighters give us too much trouble we can always just tuck tail and run to space. Grandin says that we're not space-worthy, but he's too picky; I think his definition of the word is different from ours."

"Different than yours perhaps. Let me define it for you—the hull has not yet been pressure tested."

"I see."

"The shields that the Crenum installed could seal the ship against the vacuum of space, but they're still too unreliable. If the hull springs a leak while those shields are down…we all die."

"Ouch! I get it."

"We could wear spacesuits if we have to, but none of us have practiced operating this ship from Crenum spacesuits."

"Something else that we didn't have time for."

"Exactly."

"Well, at least we can out run anything built on Earth, easily, except maybe missiles. We can go Mach 15 at sea level with those alien shields operating; they shape our aerodynamics in some way; I don't understand it completely. You designed a spacecraft, not an aircraft; without the shaped shields we are about as aerodynamic as a golf ball."

"Another advantage to conspiring with aliens," Stan said with a chuckle. "How fast do the U.S. fighters go?"

"Most of them are the same aircraft types you're already familiar with. Budgets haven't allowed for purchasing very many new fighter aircraft, so they've mostly just been enhancing the existing ones with new avionics, anti-gravity, high-G compensators, graviton scanners and jammers, etc. It makes them all far more deadly than they were when you flew in them."

"Ya, that's what I figured; I heard that most fixed-wing aircraft can now fly like hovercraft, for one thing."

"You heard right, and that's big in battle. They have also pulled a few F-4 Phantoms out of mothballs, given them new engines, particle shields, a huge anti-grav system, and stuck a plasma beam emitter on the nose instead of the huge cannon they used to have. Big seems to work better now with anti-gravity, and the Phantom's size, and big twin engines, give it more power for bigger toys. There's also the Air Force's new advanced fighter-bomber you may have heard of, the F-58. It has all those new gadgets built in instead of added-on, tops out at Mach 10, and fires particle cluster cannons. We don't have very many of those birds though—they're too expensive. They assigned those new fighters and the upgraded Phantoms to the 1st Advanced Tactical Fighter Wing, along with a bunch of other upgraded birds. It's the only Earth-based unit, short of a nuke, that matters very much against alien shields."

"Particle beam clusters, plasma beams, how come you couldn't get me any of those gadgets."

"Hey, I got you the best that could be had. That other stuff is just too new and well guarded. Seems like close control of certain weapon types is one of the clauses in that Zom treaty with our government."

"Even the very best graviton sensors, radar, and infrared aren't as good as what the aliens use, and visual sensors aren't made on this planet. Those aboard our ship are automated; they detect movement, zoom in, and identify visible targets in 3D, and they do it faster and with better accuracy than a flesh-and-blood pilot can do alone.

If the computer is having trouble with a target, it'll show us the best image it can get, so we can extrapolate. Also, alien weapons will memorize, lock, and chase a visual image.

"None of Earth's fighters can hide from us, we...they...have absolutely no stealth abilities that matter a wit to our ship's alien sensors, and we don't show up on any Earth-made radar or graviton sensors. We can see them when they can't see us."

"What about infrared?"

"Well, I was getting to that. The hull has no external heat sources, and the stealth shields match the temperature of the environment...mostly. So infrared still works, but our image is so weak that it would be more useful for collision avoidance than for weapon targeting."

The fighter pilot in Stan pondered the possibilities of flying such a craft into battle.

"So it would seem that they have to be able to see us with physical eyes or not at all. I don't think Earth has any automated systems that can track a visible image. We'll be traveling at night anyway."

"Visual motion sensors can see us, in day-light, but only close-up and if we're flying low then the moving ground makes it almost impossible to track us with Earth-built motion sensors. They just can't process information fast enough to separate us from the ground clutter. The stealth shields blur us slightly and reduce how much light reflects off us. The only way any Earth-built weapon can lock and chase on us is by painting us with a laser, and even that's iffy. We are invisible to everything else."

"Provided the shields are operating."

"Right, well they are the key to the whole system—and our Achilles Heel. The shields on our ship are standard alien shields with a stealth capability added in. All alien ships use shields, but stealth shields are special; usually only fighters, shuttles, and probes are small enough to fit inside the stealth envelope. The ability to wrap stealth shields around a transport-sized ship is new, and only the Crenum know how to do it—but on our ship, the system's gone all screwy. For us, they had to increase the size of the field quite a bit, and had to use some very experimental tricks to do it. They wanted another week or so of testing and tweaking to make sure the system's stable. If at any time those shields stop working, we become just another Earth aircraft, visible to everybody and vulnerable to anything that we can't escape from at Mach 5."

"Ok, tell me about Zom fighters."

"Stan, Zom fighters rock."

"That bad huh?"

"They have the best fighters, and pilots in UCP space. Their visual sensors are a little bit better than the Crenum ones we have, and their use of sensors in combat is unequalled anywhere. Plus, we're a lot bigger than a fighter, so we'll be more visible than they are anyway. Unless we see them first, we might never see them at all."

"Perfect. And what happens if they start shooting?"

"It would be a very one-sided affair where they do all the serving and we do all the eating, no fun at all. Plus, they can go anywhere we can go, and get there tons faster. The weapons are bad news too, plasma missiles—specially designed to disrupt alien shields and fry power systems—and that's when the shields are up. Any hit with the shields down will burn a big hole straight through the hull and kill us all."

***

On the evening of the flight, as sundown approached, they began their last-minute preparations for departure. Grandin arranged for extra people to patrol outside—it always amazed Stan how Grandin managed to move personnel in and out of their little base, and to communicate with his superiors and others both on and off planet, safely and quietly.

Under cover of the half-light of dusk, they opened the huge service door at the top of the cooling tank that had been the ship's birthplace. Stan sat at his post, strapped down in the pilot's seat, performing pre-flight tests.

He looked out through the transparent walls and ceiling of the domed bridge. The Crenum, who had been calling the cave "Mountain Base", often said that they hoped they could return and use it again someday; it was a good outpost. Stan knew that was probably unlikely. He felt a brief twinge of longing. Everything that he still possessed, that he'd get to keep and take with him into his future, was on the ship. He still had money, locked away in banks and businesses that he might have access to in the future, but everything else that he could not pick up and carry would stay behind. His comfortable house, and his hidden home away from home, would now be empty.

The ship had always taken what it needed from him. Today it would make him leave the rest.

They had hastily christened their ship "*The McPherson*". Claire suggested the name, and the Crenum unanimously adopted it through use, all against Stan's wishes. He would have preferred to call it the "First Earth Interstellar Space Exploration vessel" or F.E.I.S.E., but when he mentioned it Claire wrinkled her nose, and the aliens collectively shook their heads.

While they waited for dark, they powered up the drive system several times, so that Stan could hover the ship while the Crenum technicians conducted tests to make sure everything functioned properly. Stan rocked the ship left, and right; forward, and back, testing the controls and getting a feel for how they handled—as far as he could within the close confines of the cooling tank. The huge facility where he'd lived and worked, and which could have housed a dozen or more ships that size, suddenly felt very small.

He squirmed in his seat, eager to be off. This was his dream come true. His wish, his life's work, was about to take to the sky. Even though they were not actually leaving the planet on this flight, his blood tingled.

Claire came running onto the bridge and buckled into her seat,

activating the combat scanners.

"We have to go right now! One of Grandin's scouts just spotted a C-130 flying low outside."

"Are you sure?"

"His description was pretty specific, I wonder why they'd send troops to evict little ol' us though," Claire grinned. "I would have expected goons in suits, or a constable with a badge or something."

"Serves me right for not paying the bill, maybe someone told them we have armed space aliens in here with us."

"No doubt."

Instinctively, Stan looked out through the top of the tank at the sky, searching for the con-trails of fighter aircraft. It was a perfect desert evening outside, with the glow of the fading day highlighting the feathery clouds with red. Pilots hope and pray for beautiful flying weather like this. Right now, Stan would have preferred a good storm.

Grandin walked in. "I have recalled my people who were outside; they are all aboard. What kind of craft did you say it is?"

"A C-130; it's a military transport," Claire said. "The only reason for it to come here is to crash our party."

"That transport is likely full of assault troops," said Stan, "and there are probably jet fighters and helicopter gunships standing by elsewhere. They must have heard that you folks are here with me. We can't wait for dark."

"If we leave now we will be seen."

Stan turned to look at him, "Grandin, *you* won't be seen. You, your people, and your technology are all aboard, and we can escape if we leave now. If they have choppers with them they can park them over the top of this tank and turn this nice little sanctuary into a prison."

"They could just be on maneuvers," Grandin offered.

"Awfully interesting coincidence if they are," Claire said.

Stan lifted the ship from the floor of the tank for the last time.

"Claire, send out one of those probes I gave you and find out what we're up against."

"Already on it."

They all watched as the tiny sphere took off and flew swiftly upwards out the top of the tank. Moments later, it returned and snapped back into its docking port outside the ship.

"The probe's sensors caught the C-130 landing in a clearing a quarter of a click from here," Claire reported. "There are also three choppers with it and about two dozen fighters circling up above. I can ID at least one of them as an F-4 Phantom. No doubt one of those upgrades I told you about."

"There goes the neighborhood," Stan said.

Grandin's face went grave. "F-4s, those are the ones with…"

"Plasma beam weapons, yep," Claire confirmed. "The modernized ones are the only ones in service, and it makes sense that they would send their most advanced birds if they thought they might encounter an alien spacecraft. I really can't picture them sending just three helicopter gunships either; the others are probably out there hiding somewhere, using this beautiful scenery for cover. Their weapons can't hurt us, but we might not see them until we're right on top of them, and then they can call in the fighters. There are probably also recon troops already on the ground nearby, watching for us."

"The sooner we're under way, the better," said Stan as he pulled back on the lift controls. Out the top of the tank sprang *The McPherson*, starship in embryo, leaving its birthplace at a rush. Stan turned the ship northeast and increased speed, skimming the treetops between the foothills at the mountain's very toes, allowing the shadows of twilight to close in around him.

"The fighters are diving from altitude, in a tightening circle," Claire reported.

"They may try to box us in so the helicopters can get over us and force us to land," Stan said. He pushed the speed control and the ship lunged forward, automatically snaking from side to side to dodge the trees and other obstacles in their path.

"They've seen us," Claire said through gritted teeth. "I've got a pair of F-58s changing course to dive in front of us."

"They appear to be quite skilled at spotting us and coordinating their attack," said Grandin. "I'm impressed. Good command and control. Your country's military seems very competent."

"Thank you," said Stan. Beads of sweat formed on his forehead, the skin around his knuckles stretched thin and white as he gripped the stick, but a broad grin lit his face. He pushed the speed controls forward some more, resolved to duck under the oncoming fighters and force them to over fly. The fighters, seeing him accelerate, spun around on their anti-grav drives and dropped straight down on top of him on a parallel course. Stan looked up and could see their afterburners blaze.

"They're trying to pinch us between them and the hill tops," Claire reported.

Stan wasn't nervous; he wasn't scared. He felt a thrill go up his spine as the fighter pilot inside him awakened. Claire drew a breath and ducked as a rock overhang on a nearby cliff zipped passed overhead. Grandin looked worriedly out the dome, and at his scanner screen as the fighters closed in.

The McPherson couldn't maneuver as fast as these fighters could, but it was still more agile than anything Stan had ever flown in his life and far faster. He wondered if it wasn't just a little bit selfish, feeling so exhilarated while his friends were so tense—so he hardened up his face, so they couldn't see how much fun he was having. He had to bite the inside of his mouth to suppress the screams of joy that were welling up inside him as his ship gained speed. It was almost surreal, maneuvering dangerously around obstacles, flying along the ground at Mach 4, yet not feeling the usual G-forces that he remembered when he used to fly fighters. It was almost like using the simulator.

The danger was there, but it did not affect him. He was in his element. Projected artificial gravity was more than just a way for an anti-gravity propelled ship to fly in space. It was also a better way of flying. The fighters, using their jet engines and anti-gravity push-drives at maximum, couldn't accelerate fast enough to stay ahead of him. He slipped under them and crested the hilltops first, plunging down into the desert valley beyond.

His biggest fear for over a decade had been that his own government would try to shut him down, now those long-held worries began to vaporize. He felt the tension lift from his shoulders as he realized for the first time that from that moment on no Earth-made creation could touch him. His ship, the source of all the dreams, sweat, tears, frustration, and recent danger in his life was paying him back now with interest, outperforming the most advanced fighter aircraft of Earth's most potent air power. The adrenalin pumped through his veins, and he could hear his heart pounding in his temples. Ever since he had first flown in a fighter jet, theme park rides had held no thrill for him, but this…this was the ride of the century.

"Agh! Choppers!" shouted Claire.

They had skimmed over a low ridge and found six Apache helicopters directly in their path in the valley beyond. The gunships, caught by surprise and unable to see *The McPherson* on their instruments, could do nothing but scramble to get out of its way. Collision alarms sounded and evasive systems kicked in, banking the ship and reaching out to the Apaches with anti-gravity push buffers, nudging each of them gently aside as they shot past in a blur. Claire screamed and covered her eyes. Grandin snapped his head around to watch them fade into the distance.

Into the Dark

Escape of the Nomad

"Yyyyyaaaahooo!" bellowed Stan as he barrel rolled the ship. Then he banked right and turned straight east, up and over a last foothill, and out into the open desert, skimming over the sage at mach 8.

Once all aircraft were well outside visual range, they increased altitude as twilight vanished suddenly and darkness fell around them.

Claire grabbed the armrests of her chair to steady her hands.

"Well…that was interesting," she said in a shaky voice.

"You both did well," said Grandin. "Congratulations."

"Claire, you've been riding around in alien ships quite a lot," said Stan, "I should think you'd be used to that."

"You're insane! The Crenum don't belly slide their ships along the ground at mach 4, dodging songbirds and jack rabbits!"

"Ah! Come on! It was fun," beamed Stan. "We should go back and do it again."

"I think we had better not test their patience any further," said Grandin. "Besides, we have a schedule to keep."

"Fine," said Stan with a sigh. He put the ship on autopilot and turned his chair around to face Grandin.

"It's time I knew where we were going."

"Scotland."

"Scotland?"

"Scotland," said Grandin, "with a brief stop at some large body of water close by, so that we can test the shields under pressure of deep water. That way if we find any problems, we will have time to correct them before we get to our destination."

"We're going to *submerge* the ship?"

"Yes. Not to worry; we do it with our ships all the time."

"We'll go to Lake Powell, in Utah," said Stan, "it's a deep-canyon reservoir and it's close by."

"Why do we need to test the shields underwater?" Claire asked.

"Because we will be hiding in a deep lake in Scotland and we need to be sure they will hold up."

"Loch Ness?" said Stan and Claire together with surprise, looking at one and other and then back at Grandin.

"That is correct."

Stan smiled as he turned back to the controls, "Excellent choice," he said.

"The United Kingdom doesn't have friendly relations with the Zom like the U.S. does," Grandin explained. "The lake is very deep, murky, and very few, even within the British Government, know that we've built a base there. That is where we will finish preparing your ship for space."

They flew over Glen Canyon Dam and looked through the monitors at the water, dark against the surrounding desert.

"There's no one out on the widest portion of the lake," said Claire, studying her displays.

They flew to the middle of the lake and stopped, then dropped swiftly and quietly to the water's surface. There, they hovered for just a moment before sliding gently into the clear water.

"It's 700 feet deep here," said Claire.

"We need to go all the way to the bottom to conduct the test," said Grandin.

Stan pointed the ship downward and dove. Even with the floor pitched up 90 degrees, and the ship accelerating past 150 knots, the artificial gravity on board and the acceleration compensators worked together to make it seem like they were still hovering and level. They passed through the 700 feet of water and stopped in the dark depths while the Crenum technicians in the bottom half of the ship went over their equipment.

"No leaks" announced Grandin, looking up from his console.

"Good," said Stan, "on to Scotland." He turned the ship around and headed back up. He slowed as they broke through the surface and then climbed to 15000 feet before setting a course straight east.

Behind them, a Zom fighter followed.

The pilot was glad to be alone.

Ordinarily, fighters traveled and fought in groups of two or more, but for this mission they didn't need any such concentration. It was better to operate separately, to cover more ground in their search.

He had hovered low among the trees, hiding from the helicopters, and listening to the Telesia's communications. Then he saw the helicopters dodge something and heard their urgent transmissions, as the spacecraft they were looking for plowed right through them. Elated, he flew straight east until his sensors found the fleeing spacecraft as it climbed and accelerated.

Somehow, they were operating under stealth, interesting trick for a ship that size. All the better for him—no one else would see them and claim a share of his kill. They were both invisible to the scanners of those other Telesian fools, who would still be searching that cave back there, and the sky around it, and finding nothing. Both he and his target had slipped away without making as much as a blip on their screens.

At the lake, when they went below the surface, the pilot had stopped his fighter nearby and waited, hovering just over the water. He had nearly lost them when they suddenly reappeared and jumped back up to high altitude.

He thought about his situation. He should take them alone...and why not? The Crenum had installed stealth shields on the craft, and likely had other Crenum technology aboard too, so they might detect him if he called for backup. If he broke off pursuit to call in, he would lose them.

It wasn't necessary to contact his superiors anyway; there would be time enough to do that after he'd completed his mission. He was not to attack them where the people of this world might witness the event, or recover any of the debris. If they landed somewhere, it would probably be at some hastily-built, carefully concealed Crenum base, and he could just leave them and report its location. The Zom could come to Earth to "purge" it of all alien trespassers on behalf of Telesians, to protect them. The Telesians would see them as heroes. The Zom would be publicly welcomed on the planet and it would fall without resistance.

If the spacecraft flew to some isolated place, like over one of this world's vast oceans, he would attack and destroy them. Then his people could come along later and collect their remains. That would permanently settle this whole matter of a Telesian spacecraft. "What spacecraft?" they would say and the invasion could begin immediately.

"Technicalities, technicalities," the pilot whispered to himself. "Why not just invade the planet and stop all this stupid sneaking around?" He could have made quick work of that winged slug of a transport back there and all of its unshielded escorts. These Telesians had such confidence in their pathetic technologies!

The Zom could destroy all of this planet's primitive defenses in a day! So what if the Crenum didn't like it, what could they do? They certainly couldn't prevent the Zom from taking what was theirs. Even if the Crenum did send a force to Telis 3, it would be long over by the time they arrived, and then they would have entangled themselves in a war they could not win.

Into the Dark

Escape of the Nomad

He, and the doomed spacecraft he was following, which the Zom often called the "*Cladnic*" (which means "nothing"), were flying east over one of the main food producing areas of the planet. He leaned over and gazed at the lights below.

Soon, these fields will serve the Zom biology mines, making and packaging food for our fleets and colonies, he thought.

He looked at the map on his console and projected his course as he followed the *Cladnic.*

He thought to himself with a grin. *Before long they will reach the ocean...*

...and there they will die.

Chapter 8

Agent Clell awoke to the sound of waves against a shore and the feel of a breeze on his face.

He slowly opened his eyes and looked around without moving. It was dark, and the blind and window were both open. A salty draft ballooned the thin curtains in and out, causing the moonlight to cast dancing shadows on the opposite wall. He could hear the distant murmur of people talking, and listened intently, but couldn't quite make out what they were saying. Simple furnishings decorated the room, primitive in design, even by Telesian standards, and a television sat on a small table in the corner. He looked again—it seemed secured to the table in some fashion. He looked around until he found the door and he could just see an evacuation map taped to the back of it.

Apparently, he was lying in some kind of seaside motel—but where?

Was he free? He tested his arms and legs, to see if they were bound and they moved freely. He sat up and his head spun, forcing him to lie back down again. How long had he been out?

It didn't matter, he was free and he had a mission to complete! He looked on the nightstand and saw what he expected, one of those telephone devices that the Telesians used to communicate long distances. One of the first things they had taught him, when the secret security agency had transferred him here, was how to use the telephone network to maintain communications silence and report his status. They required him to speak English, never Zom, when making contact, and he had a coded system of metaphors common to this planet that he used. He tried to roll over to pick up the transceiver and felt a stab of pain in his side; his whole body ached as he moved and he was unbelievably weak.

What had they done to him?

He tried again to sit up, and fell back onto the pillow, squinting against the pounding headache and dizziness. Then he rolled over, more slowly this time, and dragged the telephone over onto the bed with him. He lay on his side, dialed "9", and then the numbers to his California office, but instead of the familiar buzzing sound that indicated a ringing phone on the other end; he heard a recorded female voice saying something in Spanish.

Spanish.

He sighed, he had never bothered to learn Spanish. However, he did understand enough to gather that the number he had dialed didn't work from where he was.

They had dumped him in a foreign country somewhere, probably in the Southern Hemisphere of the Americas. *Great!* He thought. He had never quite gotten the hang of international dialing either. The people of this planet maintained so many separate communications networks for these "telephones" they used, and all of those systems worked differently—it was all so stupidly archaic.

As his mind slowly cleared, he tried to remember how he had gotten into this predicament. His mission had been to get the former Telesian astronaut, Stan McPherson, to reveal where he was secretly building his illegal spacecraft. The mission was to secure the site, and craft, and arrange for its disassembly and destruction—and then to get rid of McPherson so he couldn't make any more trouble. This was supposed to be a backward world, one that had passed its peak and would never become an interstellar race. If McPherson, or anyone else, were to fly a spacecraft off this planet, then it could compromise Zom plans for the region. Those idiots at the UCP and those paranoid, self-righteous Crenum would see to it that Telesia became part of the Crenum Alliance.

He had convinced that fool McPherson to bring him right to the ship, in the bowels of an abandoned industrial facility of some kind, but that meddling ass, Colonel Grandin, had beaten him there. Their presence had surprised McPherson, and they all exchanged words.

He tried to remember what happened then. One moment McPherson had told the Crenum to leave, they complied, and Clell was working out how best to alert his office, when suddenly he had woken up here, his body feeling as if he hadn't stood on his feet in months. He closed his eyes and forced his brain to work through the fog.

He remembered Grandin shouting at McPherson, then issuing an order and his people turning to leave. Then McPherson had suddenly turned, raised his arm, and pointed at him with something in his hand.

McPherson had shot him!

He groaned—how could he have been so stupid? He remembered then, the explosion of pain and light. He reached up with his hand and felt his face, where a small bulge in his skin remained to mark the place where the projectile had entered. He had nearly gotten himself killed by a harmless scientist with a hidden weapon—not quite as harmless as he'd seemed, apparently.

Somehow, McPherson had smuggled some kind of firearm and gotten it out without Clell noticing. Grandin had seen the weapon though, curse him, and gone ahead and let Stan use it. That was why he had pretended to decide to leave! Clell remembered now, McPherson had raised the palm of that hand and showed Grandin what was in it; and Grandin and his people would have seen the tiny gun hidden there. Stan's behavior had seemed a little bit odd at the time, but Clell had expected him to get a little crazy, when he realized that his precious ship was about to go up in smoke.

McPherson had come armed and prepared to kill. He remembered feeling the bullet sink into his face and realizing the depth of his error. Clell had never expected anything like that from McPherson. He had misjudged the man badly.

He probed the scar with his fingers, the bone underneath felt smooth and completely healed. The Crenum doctors had done well, but had left the scar, as a sign to him and his own people of how stupid he'd been.

Grandin would not have dared allow Clell to die, not unless the injury had been instantly fatal. The Zom and the Crenum were technically not at war, and if his government ever discovered that Grandin had knowingly permitted a helpless Zom agent to die in his custody, it would have been difficult to explain, and Grandin would be careful to avoid that kind of trouble.

Grandin and his people knew that he couldn't speak Spanish, but the mind of the Col. Grandin that Clell knew would not have thought of this—this rented room by the sea in a foreign country. This arrangement had the workings of someone with a twisted sense of humor.

He felt in his pockets and discovered them empty, wait! His expense card was there. One of those credit card things that Telesians used for money and that Clell always carried with him. He pulled it out and looked at it. Why had the Crenum let him keep it?

No, that didn't make sense, there had to be more.

He looked around again and saw a white envelope on the edge of the nightstand. His captors must have tucked it under the phone. He took the envelope and felt it, then he sniffed it warily; nothing appeared dangerous about it. He opened it carefully and recognized Claire Saunders' handwriting. How many times had he seen it during those months he had spent tracking her down.

Nosey busybody! Four years ago she discovered records that those incompetent U.S. officials had not hidden well enough—documents that detailed some of their dealings with the Zom government. Saunders had been about to go public, and those fools in U.S. Intelligence hadn't been able to find her; she'd just melted into the population. The Zom governor had sent him to deal with her.

He had found her, and was just starting to work on her to find out where she had hidden the printouts, when she broke a wine bottle over his head and escaped. He had to track her down again, and was about to finish her off for good, when Grandin had shown up. Not only had Grandin prevented Clell from killing her, but the documents ended up in Grandin's hands.

If Claire did arrange this, then there had to be more surprises waiting. To have him just wake up in a motel in a Spanish speaking country would not have been enough for that sadistic wench!

Clell grumbled and read the first line of the letter…

Paul McCLELLan huh? Cute.

"Paul McClellan" was one of the Telesian names that he used. It was on all of the Telesian identity documents that he normally carried with him, including his passport and credit card.

Listen, we nearly maxed out your expense card, Paul. Sorry. Everybody wanted souvenirs. I hope you don't mind.

Souvenirs? Clell tried to picture a group of barely disguised Crenum soldiers and a badly dressed, but attractive, Telesian female, rummaging around in the hotel gift shop. He shook his head and smiled.

We will of course be neglecting to tell anyone where we left you; however, we will tell your people that some scientist shot you in the face. That's just too good to keep secret.

You're lucky, Clell...I mean Paul. They wouldn't let me hurt you (well, actually...I did kick you in the ribs, once or twice, when they weren't looking, but that can just be another one of our little secrets).

Clell felt his ribs. Unlike the injury on his face, which had been skillfully treated and fully healed, his ribs bore fresh and swollen bruises. He decided if he ever got his hands on Saunders again he would return her act of kindness, with interest.

We told the girl at the front desk that you were a boozer and we checked you in to sleep off a binge. We used your card and thumbprint to pay the bill. Hope to see you again soon, when I have a blaster in my hand and no babysitters.

Into the Dark

Escape of the Nomad

Blaster! Abominable! Leave it to that barbarian Saunders to use such uncivilized language to refer to a weapon!

P.S. Better be careful down there Clell, you might catch something you don't have any resistance too. Wouldn't want you ending up in the hospital or anything. They might notice your unique physiology.

He winced at the sharp pain in his head and side as he rolled over to the nightstand again, pulling open the drawer. He shuffled through it until he found a travel brochure and held it up to the dim light.

Puerto Vallarta, Mexico!

It all came at a rush. Foreign country, Spanish speaking, tourist trap, little money, with no passport…he jumped to his feet and scrambled drunkenly to the bathroom, collapsing to the floor with weakness and pain halfway there and scrambling back up to look in the mirror, his alien face looked back at him—no disguise! They had left him there with no ability to blend in, and probably set up some way for him to attract attention. He looked back at the clock on the nightstand. It had been unplugged from the wall; the display was blank.

What time was it? What day was it?

He walked over to activate the television but it didn't turn on, yet he noticed that the distant sound and voices he'd been hearing in the background stopped. He pushed the power button again and the voices returned. He stared at the screen close up and could just make out dark shapes moving on it. The T.V. had been playing in his room the whole time he'd been lying there asleep, with the brightness and volume turned down.

He looked the unit over and found the knob that adjusted the brightness and turned it. There was a movie playing.

Pay-per-view!

His brain groped for the meaning of that. Primitive as this culture was, the intersecting demands of convenience and greed had perfected the mechanisms of instant credit card charges. Even hotels in the less advanced countries did it, actually removing the funds from the account at the exact moment the customer used the service. If there was any money left in the card—his only vehicle back to where he needed to be—it had been draining out through this television at the rate of about $20 per showing, the whole time he had been unconscious in the room. It would be just like Claire to pre-order a back-to-back schedule, going out for several weeks. If Clell had just gotten up and left the motel, not noticing the TV, it might have been playing continually for days.

How long had it gone on already?

He had to inventory his resources and stop the bleeding; he might need every penny to get back to where he belonged.

He walked carefully back over to the bed and picked up the phone receiver again, hoping that someone at the front desk there could speak English.

***

Stan looked out over the lights of New York City, and breathed a sigh of relief as he watched them fall into the horizon. Though he felt ecstatic at escaping the persecution of his government, he was also saddened a little by the sight. The more he thought about the events that unfolded around him, the more he feared for his country and his world.

His ship was flying! He was seeing the unfolding of his dream faster than he had ever imagined, but in his heart, he could feel an approaching doom. He didn't hear all of the news; he just gathered information from his new friends and put it together, arranging it all in his mind a piece at a time. The aliens that he had worked with tried to be kind and hide it, "Oh, it will be fine," they always said, "once we get the ship safely out of the system," but then Stan would look in their eyes.

The home that he knew was about to sink into the past, just as New York was now sinking into the dark ocean like a fading sunset. His native culture was about to pass into history. The thought had been working its way from his heart to his brain ever since his first conversations with Clell.

The Zom were going to invade Earth.

He would probably get to complete the ship, and perhaps in the end, they might stop the Zom from destroying any other developing worlds—but not in time to save Earth. Maybe that had always been little more than a vain hope to begin with. Could he have prevented the upcoming tragedy? He thought about it and decided that he couldn't have. Few people had worked more aggressively to prevent what happened to the space industry or sacrificed more time, money, or personal safety to turn things back around. He had planned on venturing to space and returning to Earth with a wealth of scarce resources…energy, technology, advanced materials…fuel for a new renaissance of discovery. He had dreamed of becoming the pioneer of a new space age. Now the best he could hope for was to escape Earth quickly enough to delay its eventual destruction, if only a little. Of course, if he had not built the ship the Zom might have invaded Earth already.

"We have to be prepared," he said aloud.

"Huh?" said Claire.

An alarm sounded; something was wrong.

"What is it?" Stan asked Grandin over the intercom.

"Just a moment Stan," said Grandin, concern in his voice.

Stan and Claire waited in silence, and then Grandin's voice came back on the com.

"The shields are weakening," said Grandin, "we must have overloaded them on our trip underwater; now the generator is overheating."

"Can you keep it going without damaging it?" asked Stan. He dove the ship and slowed to mach 2. If the stealth systems failed, they would soon be visible to the radar and graviton sensors at JFK Airport and others. Flying low, they would be below most radar and hard to see on any graviton sensors that might be pointed at them.

"The shield generator will shut itself down in about thirty-five seconds," sighed Grandin. Stan could hear the other aliens jabbering excitedly in the background, "There is nothing we can do to stop it. It will be a few minutes before they cool down enough to restart, but the problem will still exist; they'll just overheat again, and more quickly. The generator has to be completely shut down and repaired."

"How long'll that take," Stan asked, as he leveled the ship a hundred meters over the waves.

"Quite a while," said Grandin, "we will have to find the problem before we can correct it."

Another alarm went off.

"There's a laser targeting system tracking on us," yelled Claire. "It's Zom!"

"Great!" Stan checked his screen just as the computer zoomed and enhanced the image of a sphere-shaped craft diving down on them. It had a triangle shaped drive ring, sleek and thin, like delta wings, decorated with blue flame. The hull of the craft was a transparent sphere. At the front of the drive ring were the craft's plasma missile launchers, like two black, curved tusks, almost touching near the ends. Then the craft disappeared behind a bright yellow light.

"It just fired a plasma missile!" shouted Claire.

"Laser clusters!" yelled Stan.

"Already firing," Clair replied.

Stan wheeled the ship 90 degrees to the North. He and Claire swayed a little in their seats as the artificial gravity systems struggled to compensate for the sudden maneuver.

The missile closed on them at an astonishing speed, flying straight in. Stan turned and looked out the back of the dome just as the sphere-shaped missile jerked to one side, releasing the ball of plasma it was towing. The force field bottle around the plasma burst open when it struck the shields, releasing the small hydrogen fusion reaction confined inside.

There was a thunderclap and a flash of light. The ship shuddered, the shields glowed blue and yellow for a moment, but the ship escaped damage.

It took Stan a moment to recover his eyesight.

"Wow!" said Claire.

"Did the lasers miss?"

"There was nothing to lock onto; the Phalanx system isn't designed to counter alien weapons," she said. "Doesn't matter though, alien missiles are shielded anyway."

Stan had an idea.

"Use the lasers anyway, but fire a double load of chaff too."

"The shields are down," said Grandin.

The fighter, closer now, fired again.

"Fire chaff!" shouted Stan.

He wheeled the ship ninety degrees again and plunged to just above the water.

Small rockets fired from the back of the ship and exploded, filling the air with clouds of aluminum streamers and confetti intended to fool radar-guided weapons. In this case, targeting lasers from the missile, the fighter, and *The McPherson*, together with the seven powerful lasers from the Phalanx missile defense cluster, all struck the glittering chaff and scattered, creating a bright, multi-facetted laser show. The missile lost its lock and a steam explosion erupted from the ocean as the super heated plasma plunged below the surface, instantly vaporizing a few hundred gallons of seawater.

"That shouldn't have worked," said Clair.

"We blinded it, but he won't let that happen twice," said Stan.

"My men say your Telesian type particle shields might afford us some protection, in case we are hit again," said Grandin over the intercom.

"Perfect!" Claire quipped.

Stan quickly flipped a switch on his console and a light came on. The system used well-known Earth technology and he had installed it before he met the aliens, to protect the ship from any radiation, small meteoroids, and other natural hazards that the ship would encounter in space. He quickly did the math in his head and doubted that they could offer much protection against alien plasma weapons.

"How long before I get those skin shields back?" Stan shouted into the intercom.

"Five minutes."

"In five minutes we'll all be dead!"

"He's behind us now, instead of above, painting us with active radar," yelled Claire, "he's firing again!"

"Counter measures!"

Claire fired more chaff rockets while Stan veered the ship hard to one side, accelerating to mach 5.

The missile ignored the chaff completely this time and cut the corner of Stan's turn. Searing heat engulfed the ship, a violent surge of pressure shook it, and an electromagnetic pulse, together with a jolt of high voltage electricity, cascaded through the ship's systems.

Equipment hissed and popped. Smoke filled the bridge as circuits all over the ship overloaded, but it flew on, bruised but functioning.

"We're still alive!" said Claire, coughing.

"Ya," said Stan, "it wasn't a direct hit, and the particle shields took most of it. They won't protect us again though, their fried." He hit the com.

"How are you doing?"

The com was full of static and Grandin replied. "There's a lot of smoke and burnt circuitry down here, I can't tell you more just yet."

"We can't do that again," said Claire.

"Lock weapons," said Stan.

The Zom pilot had seen them slow from the hit and grinned. His trick worked. When he saw that he couldn't hit them with laser-guided weapons, he'd switched to gravitons. After giving his ships sensors a moment to memorize his target's signature, he painted them with radar, to fool them, then launched. Sure enough, they fired off metallic debris and turned hard to one side, running smack into his missile.

Then suddenly his target changed course. With more agility than he expected, the primitive freighter suddenly slowed, painting his fighter with a variety of radar, laser, and graviton energy. How could they think their pathetic weapons could harm him? They could do nothing! He laughed as the lasers and chemical explosives splattered harmlessly against his shields, filling the sky with an impressive but harmless pyrotechnics display.

Then he stopped laughing. His enemy was gone. They had disappeared from his sensors and vanished from sight. He turned his fighter in a slow circle and adjusted his equipment. He set the computer to focus on nearby targets, and strengthened the signal—nothing.

He had failed. The target he had carefully followed for almost the length of a continent must have somehow gotten their stealth systems active again, temporarily blinded his visual sensors, and escaped.

"You fool," he sighed. He programmed his ship to follow a slow, sweeping search pattern, a cursory path that would conduct a grid-by-grid search.

Then he saw something on his sensors. A weak electromagnetic signal, the kind caused by high voltage electricity arcing across damaged circuitry. Its source was heading straight east, across the Atlantic, gaining altitude and limping along at mach two. He smiled as he cancelled the search and turned his fighter to follow it, keeping his distance. They were probably unaware that some of their battle damage was lighting up their ship for him like a beacon, giving away their location in spite of their stealth systems.

He programmed his missiles to lock onto that signal—and this time he would not miss. He would swing around in front of them and fire two missiles directly opposite their flight path. It would give his target no time to react.

They would never even know why they died.

Like a submarine, *The McPherson* lay still and quiet, waiting, with only a small antenna and visual sensor protruding a foot or two above the wave tops.

"He took the bait," said Claire looking at Stan with a grin. "He's following it east."

"Beautifully done," Grandin added over the com. "Zom fighter pilots aren't stupid, but they aren't rocket scientists either."

"Well then, the next time I catch a fish with a spinner I'll explain all that to him, maybe he won't feel so bad," Stan laughed.

"We'd best be getting out of here," said Claire, suppressing a chuckle, "before he kills your wounded minnow and shakes off the hook. Handy little tools, those probes of yours, we should make some more of them."

The McPherson rose slowly up, shedding water in a mist as it accelerated straight North, gliding just over the ocean surface.

Just as the signal from the probe was about to fade into the distance, it ended abruptly.

They didn't hear the Zom pilot cursing.

Chapter 9

The CIA director sat in his office, stewing, waiting for news. Why should the Zom care so much about a single, privately funded project, that they would pressure the U.S. Government into taking such heavy-handed action? They were starting to look a lot less like friends and a lot more like paranoid dictators, but that wasn't as bad as the United States military receiving orders from the President to launch this kind of operation against one of its own.

Not only might this be the end of the McPherson project, but it also looked like a new escalation in Zom policy. At least President Atequipa had insisted that the rules of engagement remain benign—*no use of deadly force unless fired upon.*

Still, the director would remember. Long ago, he'd learned to stow his anger, and to bring it back later as a catalyst for action.

He credited this one in the Zom's' account.

Someone knocked at the door.

"Come in!" he barked.

Agent Mills walked in with a stack of photographs. He was beaming, but closed the door before he spoke.

"He got away."

The director stood up and creased his brow. "What?"

"Damned if he didn't make it! McPherson slipped away right under our boot heel. One minute they confirmed the ship was still in the hanger, the next minute it just lifted off and flew away. They couldn't see it on their instruments, and it was fast—oh, way to fast for anything we have. It left as soon as our infantry landed, shrugged off our air cover and vanished into the night."

He slapped his desk and grinned.

"It nearly ran over a few of our Apaches on the way out, and then poof…gone." Mills made a little exploding motion, followed by his hands spreading apart, fingers flailing for effect.

They both laughed as the tension drained away.

Mills handed him the photos and he thumbed through them. There it was, just like in the NASA designs, only real—but the shots were dark and blurry.

"It must have been moving very fast," he said. "They don't seem to have caught any details."

"They said it isn't just the photos. The ship *looked* smudged. See…they took this shot from the belly camera of an F58, during a moment when they matched speeds—it shouldn't be blurry. That camera is designed to shoot fast moving objects anyway. They said it looked like the UFO pictures that you see in the tabloids. The rare glimpses we get of Zom spacecraft in flight look like that too."

"Interesting."

"Where do you think they went?" asked Agent Mills.

"Who knows? At least now if anybody asks us about it, we can honestly say, DUH!" The director twisted his mouth mischievously. "My guess is they won't even try to hide stateside again. Probably headed to some foreign country somewhere or maybe even…" he pointed his finger straight up toward the ceiling.

"That's too much to hope for," Mills said. "At least if they went to France, or England, or somewhere like that, they won't have very many of their fellow Earthlings chasing them."

"I don't think you or I ought to start speculating about *exactly* where they went, someone might ask us to help find them," the director said. "Let's both go home now and get some sleep. They're more than likely hidden and safe by now."

***

"The main reactor just shut itself down," Stan said, looking down at his displays. "The computer says the fuel's contaminated."

Claire looked at her console, examining the list of compounds.

"Looks like seawater," she said.

She entered commands to switch to the auxiliary tank and restart the reactor.

"What now?" Claire asked.

"Well, it's not a big problem yet. We can still get where we're going, if nothing else goes wrong." He keyed the intercom, "Colonel, how long before the shields are back up?"

"We found the problem, about 15 more minutes."

"Could you send someone forward to check for battle damage? It looks like the main Helium 3 tank took on some water during our little dip in the Atlantic."

"I'll send myself."

"It took me over a year to scrounge enough helium 3 to fill that main tank, and it took just one night to blow it all."

"And it will take us an hour to replace it, once we get to Loch Ness Base," said Grandin.

As they approached the coast of Greenland, Stan turned east, to stay out of the coastal radar.

"Mr. McPherson," it was Grandin on the com, his voice seemed muffled by something. Stan and Claire exchanged worried looks.

"Yes?"

"What is our propulsion status?"

"The controls are a little sluggish," said Stan, "but I can't tell you much else. My monitors for that system are toast."

"I am in the middle, front portion of the ship and I am seeing some shrapnel damage clear through the drive ring. There is water and ice everywhere."

"Any other good news?"

"Ya, it looks like the forward four segments of the drive ring are all ripped up, and it looks like the damage disabled the PAGA gravity well, so don't go too high or too fast."

Stan creased his brow. If the PAGA projector was down, then the anti-gravity that kept them aloft had to push off of the Earth's surface instead of the artificial gravity well that the PAGA drive provided, which meant that maintaining higher speed and altitude cost a lot more power. Their anti-grav push system was part of the PAGA drive ring, but its design was simpler and more robust than the PAGA projector.

"You might want to send some men to inspect the links to the drive ring; I need to know how many circuits are blown."

Power distribution to the drive ring was shared among ninety-six separate conduits, eight for each of the twelve segments of the drive. Trouble was, as some circuits failed, it added extra strain on the others, and the plasma missile hit had already done significant power system damage.

"I have already put some people on it, but we are a little short down here at the moment. We took casualties in the fight and I have a full repair crew working on the shield generator."

"Stan, dear," Claire said, "as navigator, I should remind you that we are going to Loch Ness to hide—it won't work if we fly over Scotland in broad daylight."

Stan checked the navigation computer. At the speed they were flying, they would be pushing the pre-sunrise.

Trying to hide and wait out the daylight would risk discovery, unless they hid underwater, and they needed the shields working for that. If they submerged for very long without them, with all that hull damage, they might never come up again.

Thirty minutes passed, with Stan flying as fast as he dared. Then Grandin came back on the intercom.

"Stan, the shields are ready to test."

They activated the shields and spent several minutes checking them. They were functioning, sealing the ship again, and hiding them from radar, infrared, and graviton sensors.

"That should hold us long enough to get to Loch Ness Base," Grandin said.

"Good, Grandin, what would you say if I told you that they might have to work long enough for us to hide in the North Sea for about twelve hours?" Stan asked.

"I would tell you to grow gills."

"But I thought that we were going to hide underwater anyway," said Claire.

"Not for that long. After we arrive they will install a dome over us and pump it dry."

"Cool, like a glass bottomed boat."

"No," said Grandin, "a transparent dome would give off too much light. We're going there to hide."

"Oh, pooh," Claire teased.

"In that case, Grandin, we need to put on some speed, but..."

"I know; the circuit inspection is nearly complete."

"What happens if we fly too fast?" Claire asked.

"We've lost some circuits because of the plasma hit, and the water," said Stan. "If we fly too fast or too high, then the circuits that are left could be strained to blow too."

"Well, we have to land while it's still dark, and the rate we're flying we might not make it there in time."

Grandin went off the intercom for a minute, and then came back, "Stan, I think we can risk mach 1.7."

"Ok," said Stan.

Stan could see the lights along the coast and turned the ship south, climbing as they approached the shoreline.

The ship shuddered.

Grandin's voice came over the intercom, "What did you just do? We heard some fuses blow down here."

"We climbed to clear Inverness; I don't want to be spotted over morning tea."

There was a moment while Grandin consulted with his people.

"We'll have to risk it, at mach 1.5, but remember our shields will soften a hard water landing, but if we crash on land we're all finished."

"I know."

The ship shuddered again and lost speed.

"Down Stan, now!" Grandin commanded.

Stan lowered to 200 feet, slowing to just under mach 1, and followed the contours of the river valley that led to Loch Ness. He looked up—some of the stars had disappeared as the black of night began to surrender to the coming day. A slight glow lit the eastern sky.

"Slow us to 900 kph," Grandin said over the intercom. They could hear the angry snap of circuits popping in the background. "A drive segment just went down."

"The sun's starting to come up, if we fly much slower they'll think they have a new monster in their lake," Claire said.

They flew on until they could just see the river widen beneath them; the ship pitched and tossed like a drunken sailor.

"Can't you fly a little straighter Stan? We have pumped out most of the water, but moisture is still sloshing around a little up front and blowing systems and circuits."

"We must've lost another drive segment or two," Stan replied, "I'm having a devil of a time keeping her lined up."

Grandin paused a moment, "Yes, you are correct. Segments 2 and 7 are now out. We are down to just five."

Stan slowed to 500 kph and dropped to 10 meters over Loch Dochfour. He could see Lochend up ahead. "Get ready to land hard," he told Grandin over the intercom.

Claire and Stan both buckled themselves in as they flew over Lochend. They could see the lights of the ferry landing. Hard working locals, preparing for their day, would never hear them pass. The shields, shaped to pass through the air smoothly and muffle all sound, made them fly as quiet as a giant owl.

"We need to stay up a little bit longer and get past Lairgmore—we don't want folks to hear us slapping the water." Stan said.

"Ok, but don't take too long doing it!" Claire replied.

Suddenly the ship lurched and they slowed again.

"There goes three more drive segments, we are now flying on just two," Grandin said.

"A gull with two feathers," Claire said, her voice shaking.

"A rock...actually," Stan corrected.

The ship fell on a ballistic path to the lake surface. Skipping off the water, it flipped end over end and, like a breaching whale, back-flopped into the water with a tremendous splash.

***

"He what? He let them escape?" The Zom governor of the Telesian territory pounded the mahogany desk with his fist. The aid glanced out the window at their view of the north side of Brooklyn from fifteenth floors up.

"How did they get away from him?"

"They ghosted him and then hid in the ocean until he left. They were last seen headed East, and the ghost drone they launched resumed that heading. He says he scanned the area later and found some traces of Helium 3, but no ship. He thinks they hid underwater there before escaping."

"They probably had a jolly laugh as they watched our fighter chase after that drone! Tell me, did it occur to this pilot at some point to call for help? Didn't we send an entire squadron?"

"He did call in, but not until after they were no longer in the area."

"He should have called in as soon as they were spotted! Those fighter pilots think they're invincible! We should have sent shuttles full of soldiers, instead of these glory hounds; real soldiers would have called in course and heading the moment they spotted the craft. Then we could have dealt with it properly and this whole embarrassing mess would be over."

"They had orders to maintain communications silence…"

"They had orders to find and destroy the *Cladnic*! It can't outrun our fighters. What difference would it make if they discovered they were being followed? None! No, this fighter pilot wanted to kill them himself and get a medal!" Then he seemed to remember something. "Where's that fool agent that was supposed to solve this problem for us a month and a half ago?"

"We heard that he'd been injured and we traced his expense account. We found him in a rented room in Mexico."

"Mexico? What in stars is he doing there?"

"Apparently, the Crenum dumped him there, Sir."

"Grandin didn't do that," the governor shook his head, "he's too straight forward for that kind of nonsense. He would have just dropped the idiot in a nice clean ditch or trash bin somewhere."

"We called Agent Clell in his room; he says it was that Telesian woman…Saunders."

The Zom commander thought about that for a moment, and then lifted his face to the ceiling and laughed long and hard.

"Well you'd better send someone out to pick him up," he said finally with a chuckle. "She probably left him without food, money, or disguise, and with more than a few bruises."

"That she did, Sir," the aid said, grinning. He was glad to see the governor back in a good mood.

"What are his injuries?"

"He was shot in the face with a small handgun that McPherson had with him."

"Clell admitted this to you?"

"No, that came from an anonymous tip that some Telesian sent us by vid-com. It was a female in disguise, probably Saunders. She even provided us with photos of the injuries, to be sure we knew what had happened. We were unable to trace the call."

"No surprises there, knowing Saunders. It had to be her of course; no other Telesian in this country would have had the means to call us on a vid-com. Have Agent Clell and that thickheaded fighter pilot report to me, together, as soon as they both come in. Forward those photos to me."

"Yes Sir."

***

"I'm not going to ask how you got your probe close enough to catch any of this," the General Advocate said reproachfully.

"I'm prepared to tell you exactly how that probe came to be there. What's more, the individuals who collected this are not only available to testify, but will soon be banging down your door insisting on it. You may use any of this probe data you like, in any way necessary, and there were no Crenum agents involved in its collection."

Steck's eyes sparkled as she looked up suddenly from the screen. "Someone else shot this vid, not the Crenum?"

"Well…not entirely without our help. We…arranged it. A science organization unaffiliated with the Crenum owns that probe. It's Kreshan."

Lertis' eyes widened,

"Really? What kind?

"One of their new Gertin series stealth probes."

"I'm surprised the Kreshans involved themselves like that," she said. "They consider themselves above politics, and they're friendlier toward the Zom than they are toward you." She fiddled with the display controls, isolating and enhancing numeric readouts from the probes sensors. "Who's the unfortunate target here, not our Telesian spacecraft?"

"We believe it is, but the incident did not occur close enough to the probe for a positive ID. Those are Zom plasma missiles however, that much is clear."

"...being used illegally inside the atmosphere of a primitive world," Lertis finished with an icy tonc.

"The Kreshans of course were incensed, they immediately cancelled all scientific cooperation with the Zom and said they'll file charges, in addition to sharing these images with us."

"How did you do this? The Kreshans are mostly scientists, not crusaders. I'm familiar with this probe design; it's their top model, very expensive, and they only have five of them."

Klicktah looked at her in amazement.

"What? Don't look so surprised ambassador," Lertis said, replaying the vid sequence again, some irritation in her voice. "I have to know a little science to do my job."

"We offered them an opportunity to photograph the McPherson craft and they sent this probe," Klicktah stammered. "As you know, the Zom let the Kreshans entre their space to work with them on the study of a red giant star there that's collapsing."

"The Fatalode star is a very long way from Telesia, Ambassador," Lertis pointed out with more admiration than reproach.

"Of course it is, but they are scientists after all," Klicktah said, grinning.

"Can't say as I blame them, cultural science has become somewhat of a hobby for me too of late—fascinating field," she said, with a hint of sarcasm. "These images are very recent, Ambassador, only an intelligence service like yours could have gotten them here so quickly."

"Your right, this is my copy, direct from the Telis system. The probe downloaded its data to one of our probes, as part of our arrangement, before it left the system. The official copy will be crossing your desk later through diplomatic channels. The Kreshans sent it here through their UCP ambassador's office, you'll be hearing from him soon."

The General Advocate raised her eyebrows, "That ought to be a fun meeting," she said. Then she composed herself, sat back in her chair, and looked hard at Klicktah.

"Now, unofficially, what happened?"

"Well, we found out from our people in Telesia that they planned to transfer the McPherson craft to a hidden base of ours elsewhere on the planet, to escape capture by the Telesian authorities. We decided to use that flight as an opportunity to obtain the independent images you wanted. So, we leaked the flight plan to a few Kreshan scientists and they rerouted one of their probes to get an image of it for their cultural history databases."

Lertis smiled again, "Of course, no Kreshan could pass that up."

"No, and they agreed to share their images with us as part of the deal."

"I should think they would have been willing to agree to a great deal more than that. You offered them an opportunity to get a vid of the maiden voyage of a grounded culture's first interstellar spacecraft. Most any Kreshan would sell his mother into slavery for such an exclusive; a very clever idea, Klicktah, yours?"

"No, such things are not my job. Director Gimlar is the one with the Kreshan contacts; he arranged it all."

Steck Lertis smiled again, "I met your Director Gimlar a few years back when he worked for Crenum intelligence, I should have recognized his hand in this. It was not just his contacts either, this kind of operating shows his particular style."

"Ya, but something went wrong, the Telesian craft didn't make it to the arranged rendezvous, the probe had to look around to find them. That's when it picked up a plasma missile detonation and zoomed in on it."

"Well, I know the Kreshans were not involved in tipping the Zom off. No Kreshan would knowingly say or do anything that could lead to the harm of a primitive culture; besides, if they'd told the Zom where the craft would be, there would have been a lot more than just a fighter or armed shuttle there to greet them—and they wouldn't have allowed this probe to escape either." She looked closely at the image viewer and frowned. "It almost looks as if the missiles destroyed their target. Where is the McPherson craft now, Klicktah?"

"I don't know. No one has told me whether or not they arrived at their destination."

The General Advocate looked grim, "Well I hate to say this, but while you failed to get the images you were after, you may have gotten something far more valuable, I only hope that we didn't lose the prototype. We can't specifically see from this data who the target is, but we can surmise it from the technology level of the drive signature. We can't see the attacker either, but those are plainly Zom missiles; their configuration and power curve are unmistakable. I'll add this to the evidence and look forward to my meeting with the Kreshan ambassador. Will you be there?"

Klicktah shook his head, "No, I don't think so. The Kreshan government is not at all happy with us over this. We set them up, and they resent us getting them involved in our disagreements with the Zom. They were deeply invested in the research on the Fatalode star and now that project is scrap."

"Well, their attitude toward interference is practically a religion with them. The Fatalode project is important, but as soon as their people learn of this incident, they're going to be hopping mad. To them this is, at the very least, a brazen use of advanced weapons in the atmosphere of a primitive world. They'll probably figure out on their own that the Zom attacked the McPherson craft, and for the Kreshans, that's explosive. It'll look to them like the Zom tried to interfere with a test flight."

"There's more," Klicktah pointed out, "the Telesians have their own sensor probes in orbit around the planet, to look for nuclear chain reaction missiles that were recently banned by the governments of that world. Their probes are sensitive enough to have seen this attack, and their people are advanced enough to identify some of the technologies used. The United States has some plasma and particle beam weapons, but no plasma missiles…yet. However, now that they've seen them in action…"

"Well, let's not tell the Kreshan ambassador that, his head will explode," Lertis said grimly. "Of course what will happen now is that as soon as all of this hits the news, the Zom will stop allowing the Kreshans into their space from here on out. You backed the Kreshans into a policy corner and they'll take flack from their people no matter what they do. They have no winning position; half of the folks will scream about the sacrifice of their precious Fatalode project, while the other half will ask why they were so tight with scum like the Zom in the first place. You've managed to drag them over to your side by their hair—no wonder they're unhappy with you."

"Too bad for them," Klicktah grinned.

She stood and handed back his data viewer. "Please give Director Gimlar my regards."

"I will, Madam Advocate."

Chapter 10

"I think you need to work on landings," Claire said.

Stan was hanging upside down in his seat, held in by his restraining straps. Clair's long red hair hung down over her head.

All the Earth-built systems, including the artificial gravity that kept the bridge always feeling level, were down. The status lights and displays of alien equipment lit the bridge with an eerie glow. There was a dim greenish light from outside as the early sunrise filtered down to them through the lake water.

Stan sighed, "Are you all right?"

"Compared to what, a crash test dummy?" Claire said testily. "I'm ok, you?"

"Shoot me, just shoot me now please."

They both laughed.

"Now what?" Claire asked.

"Well, we hang here like bats and wait for the green guys. They're up; we're down. That kind of puts them on the rescuer side of things and us on the 'help me help me' side." He said the "help me, help me" part in a squeaky, high-pitched voice.

"If I ever meet that fighter pilot I swear I'll scratch his eyes out."

"Wait in line."

"So, Stan, were you ever married?"

Now where did that come from? Was he ever married? He'd gotten excited about space flight as a child, following the X-Prize competitions on T.V. and on the Internet. He completed school at the top of his class, and flew fighter and cargo jets in the Air Force for exactly the number of hours necessary to get into NASA. He was a pilot and an astronaut but he had always been a scientist first. After a couple of space flights, he had transferred quickly to the propulsion division just as things started getting interesting, and was instrumental in the final stages of the anti-gravity development project. He had been with the Newton Gravity Science Lab at NASA for five years when the unfortunate truth about anti-gravity had hit.

Then, someone had given him hope. A project group that was working on the PAGA concept recruited him as their spokesperson. The "catapult to the stars" they called it. In the end, that project only proved two things—first, that the idea of a PAGA drive was mathematically sound; second, that even a fast moving, successful endeavor can be tossed in the trash if nobody cares. The PAGA drive project became a metaphor for his spaceflight science career—flameout and fizzle.

Embittered, and soon to be unemployed, and with a wealth of skills, knowledge, and experience that were now useless, he was deciding between re-joining the military or flying jetliners when he had an unexpected visit from a sympathetic friend in the CIA. This friend, Gastone, a.k.a. Agent Mills, told him that there was a government planned "accident" that would wipe out all of the confidential research data at the Newton research park, including the database on the PAGA drive project. Having access to that data, and having more knowledge about computer security than he should, Stan and his friend had no trouble sneaking into the facility and stealing the files.

After that, he sold his house and rented an apartment in the country. He lived off his savings and studied whatever details in the data that he had not already known. He secretly did many experiments that would not be duplicated, and made countless discoveries that would not be published.

Then his father died, leaving him land, a mansion, and many business interests to manage, both small and large. He sold off most of his properties, keeping only those that he could use for his plans. For several years he had a big house, a big pool, a big cave, a lot of money, and many corporate holdings and contacts from his father's variety of business ventures through which he could obtain equipment and materials in ways that were not easy to track.

Having a new purpose in life, he'd jumped into the project with both feet, postponing the things that normal men did with their lives—and a decade and a half goes by in a hurry when your hands are full.

Now his ship, his equipment, his data, everything he had sacrificed a family and every other good thing in life for, was adrift, wrong side up, in a very large, very cold lake half a world away.

"No," Stan sighed, "I've never been married, this is my family."

"Wow, bummer. Well, there's a bright side to this."

Stan laughed, "Really? What's that?"

"No one can find us."

He looked through the dome of the bridge and could see water insects slide past as the ship slowly sank into the murky depths of the lake. A dark eel and several fish glided by and disappeared into the gloom. It was growing darker and colder, in spite of the rising sun outside, and the water was slowly turning from green to brown—dark and foreboding.

"I remember back when you were with NASA," said Claire. "What happened to all those beautiful women who like to ogle famous people? I should think you would have had time to hook up with one of them."

"Well, I dated a little bit, but nothing ever really worked out with any of them, family-wise."

"Well, I guess you missed your chance now. You're not famous anymore."

"Nope, and my life is kind of upside-down right now anyway."

Clair giggled.

"What'll happen if this doesn't work out? I'd think you could find something else to do with yourself. Knowledge is a powerful thing."

"Claire, if there were a space propulsion scientist standing on your front porch, and you wanted him to go away, how would you do it?"

"Who says I'd want him to go away?" She looked at him and winked.

"It's a riddle, Claire."

"Oh, sorry. Ok, how would I get a space propulsion scientist off my front porch? I don't know, maybe sic the dog on him?"

"Nope, just pay him for the pizza."

"Oh come on, smart people are always in demand."

"Not me. Even if I was, I'm addicted to space. For fourteen years, I've been slowly going out of my mind. You want to know why I never married? It's because I think nothing should be more important to a man than his family, and all my life space has been the only thing I've ever really craved."

"Is it still?"

"Now more than ever."

She went silent, and he realized he might have answered that last question a little too quickly.

The intercom beeped.

"Hello, Stan? Claire?"

With an effort, Stan reached up and pressed the button.

"Grandin, good to hear you weathered the crash. I see you got the com working," Stan said in an overly enthusiastic tone.

"Ya, and not much else."

"Shields?"

"They are holding, for now anyway. The relevant systems all seem to have done their jobs, just before most of them went out for lack of power. The Crenum systems all still work, but those are not the ones we need right now, except the shields, of course."

"Artificial gravity is out up here on the bridge."

"It is out down here too."

"Well it may still be functional," Stan said, "just no power."

"Your right, are you two ok for a while? We need to complete our damage inspection and try and get some things working down here."

"Ya, I think better like this anyway—I read somewhere that it improves blood flow to the brain. You should know that we're sinking, and that it's starting to get cold."

"Ya, we know. We will try and get some propulsion going, but if we hit the bottom there is no danger, as long as the shields hold."

Claire nodded, "Ya, and if they fail we'll all drown anyway."

"As for the environmental systems," Grandin continued, "we will work on those first, just as soon as we are sure the ship is out of danger."

It took a while, but the Crenum technicians managed to get some of the power restored. The bridge started to warm up, and some propulsion and flight instruments activated. Stan turned the lights back on and righted the ship, then everyone went to work resetting breakers and replacing fuses.

"Well we are back in business," Grandin finally said with a sigh, as he joined Claire and Stan on the bridge. "I do not think we can risk flying again for awhile though, the drive ring is going to need extensive repairs,"

"I guess we'll just have to be a water creature for now," Stan said with a slight sarcastic edge to his voice. "Let's get going, are you monitoring those shields?"

"No problems yet, but we should try and find the base as soon as possible. We are late, and we have to get this ship under a bubble while the shields still hold."

It was about noon and none of them had had any sleep; it had been a short night, but a very long morning and Stan was eager to find a place of safety and get some rest. The earlier failure of the shields, combined with the leaks, motivated him to stay awake. If the shields went out again, the ship would probably fill with water, lose power again, and sink to the bottom. They all got out their spacesuits and kept them handy, just in case.

Visibility was nil and the navigation computer was down, so they groped their way around the lake, using low-power sonar and a map that Grandin had brought with him. Once they arrived at the general location of the base, they wandered for thirty minutes back and forth, the passage of their ship churning up centuries of silt, before nearly colliding with the wall of a tower.

"There it is," Grandin said.

The structure consisted of Crenum transport ships, docked together like slices of pizza, and stacked on top of one and other. When Stan had first seen a Crenum transport, he'd wondered why it was shaped like a triangle, with a slightly rounded stern. The shape had seemed odd to him at the time. As they ascended the "wall", Stan could see that the ships had been stacked five deep like the layers of a great wedding cake.

"Welcome to Loch Ness Base," said Grandin.

"Cool!" said Claire.

"Yes, it is one of our smaller enviro-modules that we use on dead planets, adapted to deep water. We attached some of the ships to the floor of the lake with heavy anchors and guide wires, to keep it steady. There are three more layers of ships lying beneath the mud, and a much larger anchor with spread feet underneath them. We build most of our environments out of our science ships like this. Each one connects to the ones next to it, both sides, above, and below. Various differences in the internal design of each ship allow them to specialize at particular functions when assembled into a collective like this.

"Like army ants." Claire said.

"What?"

"Army ants are an insect in Africa," Stan clarified. "They travel through the jungle and when it comes time to rest, they link their bodies together to form a shelter."

"Where do we fit?" Claire asked as they reached the top of the tower. Stan banked *The McPherson*, so that they could see the top of the tower clearly. There were lights shining out through the transparent forward sections of the alien transports, creating a single large light, which shined upward from the center of the tower, illuminating the cloudy water like a ghostly beacon. Looking down into the light, they could see the station's crew moving and working inside their ships. Occasionally, someone would look up and point at them.

Grandin put his hand on Stan's shoulder, "They know we are here, land on the top of the collective, in the very center, please. The base personnel are waiting to install the dome over us in segments as soon as we touch down."

Like a giant dragonfly landing on a rock, *The McPherson* spread out its feet and perched lightly on top of the ships that made up the highest floor of the tower. The structure itself was barely as wide as *The McPherson* and Stan could see that each layer consisted of twelve ships.

Then alien shuttles came, each bearing one of twelve interlocking segments of a huge half-spherical dome, and arrayed them around the outer edges of the tower until they had completely enclosed *The McPherson.* Once all were in place, streams of bubbles rose from the alien ships below, until an expanding pocket of air appeared in the top of the dome.

The rest of the aliens that had flown with them came up to the bridge, some carrying wounded comrades on stretchers. They all waited and watched as the air pocket enlarged and began encompassing the large bridge view port so that they could see the inside of the dome clearly.

They heard a rhythmic thumping coming from under the ship.

"They are sealing off the battle damage from the outside using force fields," Grandin said suddenly, putting away a communications device that he had been using. "In a minute, they will attach a boarding tube to the docking collar. Will you join us aboard the station?"

"I'm not leaving until my ship is high and dry," said Stan.

"There is nothing left for any of us to do here except watch them pump the water out."

"I understand that," said Stan, "I'll stay and watch. Also, before you go, I need to talk to you privately."

"Of course, let me see my people safely off first."

Later, they met in Stan's quarters aboard ship. Lines of bubbles were gliding along his partially submerged view port. It reminded him of an aquarium.

"I wanted to go over what happened out there last night," Stan said, he deliberately lined his voice with a little iron. He wanted Grandin to understand how upset he was.

"Yes?" The colonel was used to giving orders and clearly did not like McPherson's tone.

"You bamboozled me."

"Bamboozled?"

"Sorry," Grandin acted so human that Stan sometimes forgot that English was a second language to him, "'Bamboozled' means 'deceived'. You were not completely honest with me about the challenges we would face to get here alive. Now that I look back on the flight, it seems that we were unprepared, and only made it here through an extraordinary burst of luck."

"Oh, that," Grandin said. "Well, there are several things to consider. First, while we did take some precautions against a possible encounter with a Zom ship, I did not actually expect to have to fight with one. Last night that pilot displayed a shocking lack of respect for our laws by firing on us with plasma missiles. His willingness to use alien weapons inside the atmosphere of your planet was far more brazen than I had expected, even from a Zom. I had not even expected them to find us at all, and I honestly did not think that we would survive such an encounter if they did. That is why we worked so hard to conceal the ship with stealth shields. That fighter should have destroyed us. We had little choice in the matter, and in a situation like that you fight with what you have."

"Now you tell me these things; you see? That's what I'm talking about. You insisted that I captain the ship, yet you withheld information from me, information that was vital to my duties as captain and to the survival of the ship and crew. Claire told me something about the danger of Zom fighters, but all of her research is arrived at independently, without your guidance. Am I correct?"

"The crew are my people; I am responsible for their lives and safety," interrupted Grandin, his voice rising, "but I am also responsible for making sure that a bad situation is not made worse by telling you too much about the technology of other worlds!"

"When you made me captain of this ship, that role was reversed!" Stan shouted back. "When I'm in command, I am the one responsible for our safety! I needed to know our odds of survival against an enemy ship. I also needed to know more about the risks from those *experimental* shields before takeoff—specific, technical risks—not just the vague, second-hand knowledge that Claire, smart as she is, provided me. From now on, you will stop regarding me as a primitive alien species from whom you can withhold information. The first time I saw Claire Saunders, she was dressed like one of you, and wearing one of your side arms."

"Well, actually, she stole that..."

Stan suppressed a chuckle. "But you let her keep it!" he went on. "Why? If protecting Telesians from your technology is so damned important, then why do you allow her that kind of access to your weapons?"

"She is an outcast from your world, almost as alien now to your people as I am. She is not in a position right now to share what she finds out about us in any significant way. In fact, she may have already told you that she considers herself more Crenum than Telesian."

"She did say that, yes."

"In fact, given what I know of your personal history, you have been isolated from your culture...a grounded man of space...for some time..."

At that, Stan nodded slowly. Grandin paused, as if suddenly realizing what he had just said.

"I see your point," Grandin continued.

"Good. From here on out you are to regard both Claire and myself as being as you say, as disconnected and alien to this planet as you are. I'm in command when this ship is in flight, and that means that you are to keep me *fully* apprised of *all* issues that might affect flight safety and ship security, regardless of how alien that information may be to Telesia. Is that clear, Colonel Grandin?"

"It is clear, Captain McPherson."

"Good, what is the plan now?"

"Well," Grandin began, "of course we must make repairs and refuel the ship quickly, in case we have to leave in a hurry again."

Stan nodded.

"Then we have to finish construction and get you ready for space. The area inside this dome can be completely evacuated of air, so we can pressure test the hull properly. We can also form the shields around the docking tube, to test them over long periods of time."

"So that if we encounter a Zom fighter, like we did on that last flight, we'll be better prepared?"

"Look, Stan. You know we had to leave, you saw those troops coming. The ship was not complete…"

"What about more advanced weapons?"

"McPherson, we have already been through that. We can't bring alien weapons onto your planet. Our side arms are bad enough, but transporting Crenum anti-aircraft and anti-spacecraft weapons planet-side is illegal in so many ways that it is not negotiable. My orders are very specific in that regard, and the reasons for those laws are compelling and the consequences for disobeying them are severe. We are here to help you, but my first duty is to my government, and they are watching this operation very closely. I will not commit treason for you, no matter how enthusiastic I am about my mission here or your cause. Our next stop after this one will be Telis 6…"

"Saturn?"

"…correct, we have a very large base there, and once we get you there we can arm you properly. By that time, this vessel will have traveled far enough, fast enough, to be considered a starship under our laws. It changes everything. You will no longer be considered grounded, and no one from your planet has the ability to access the place where those weapons are kept."

"Saturn is some distance away, what if we get in another fight en-route? If we have a hull breach like the one we had last night, it will go far worse for us than it did over the Atlantic. There is no ocean to hide in, and we need the ability to defend ourselves better than we did. You say that it's illegal for you to arm this ship properly, and I understand your reasons, but the Zom don't seem to care much about that. You're here helping me, under questionable legal circumstances, because of the Zom, and you and your men all carry weapons foreign to this planet. I remind you again that Claire carries an alien pistol herself."

"And I remind you again that she stole it."

"But why can't the same principles apply to weapons mounted on the ship? We needed them last night. We were sitting ducks and the next time this ship flies, we may be again. Last night we flew roughly seven thousand kilometers and we almost didn't make it. When next we launch, we will be flying over fourteen Astronomical Units, well over two-billion kilometers, through the vacuum of space. It takes time to drown in the ocean, Grandin, but in space, death is instant. You must understand my concerns here."

"All too well Stan, but this ship is a very small spec in space, and 2 billion kilometers is an awfully large area for the Zom to search. What you do not understand is that we have helped you to build a very experimental spacecraft, none of this has ever been done like this before, and what we did last night was a test flight. This ship is experimental, built under experimental circumstances, with a purely experimental combination of your technology and ours. Our risks out there last night were mostly a product of the fact that everything about this operation is an experiment, and if we are standing here, with breath to discuss it, then the test was successful and we should celebrate. We will celebrate, later tonight on the station.

"Now, there were factors which we planned for, and others we did not anticipate. Yet we made it, and we learned. With or without weapons, we will still be in much better shape for our next flight, because it will not be our first. Think of what you have just done. You have flown this craft through two different environments, under fire. Is that not the most accomplished test flight in the history of your planet?"

"I crashed!"

"We lived!"

"We almost didn't. The first submarine ever used in war in my country sank several times and taught it's makers a lot of lessons, but in the process it killed more of its own troops than it ever did it's enemies. We can't afford to repeat those kinds of successes, Grandin."

"True, and I promise to give you a more accurate assessment of the odds next time. Look! They've nearly finished pumping the dome out. When the water is gone, this ship will no longer be in flight status and I will resume command. My first order will be for you to board the station at once, to relax, and to celebrate. Why don't you and I secure the ship and get ready to join the rest of the crew; there is another Telesian on the station who I would like you to meet."

"Another Earthling?"

"Yes, and he has worn out our ears nagging us for an opportunity to shake your hand."

Stan lowered the ramp of the ship and he and Grandin stepped out onto the still wet floor of their new "hanger". Some Crenum technicians had already begun assessing the damage and unloading equipment. Claire was waiting for them, standing a ways off talking to several aliens, and to a short, stout, dark haired man with a bushy mustache.

"Good to see you again, Grandin!" the man boomed as he strolled forward. "I hope your time in America hasn't corrupted your English." He spoke with a thick British accent and walked with a jovial gate. He grinned and winked as he seized Stan's hand in a bone-crushing grip. "And Doctor Stan McPherson…or should I just call you Dillinga'?"

Stan took a moment to grasp this. With all the recent excitement, he had almost forgotten about his little incident with Clell. Of course such news would travel fast enough to reach here by now.

"Stan," Grandin motioned toward the man, "this is Agent Jack Standish, British SIS."

Stan recoiled—and almost wanted his hand back. Here was a spy of an Earth government, a government that traditionally worked very closely with the U.S. government from whom they had just fled.

Jack looked sideways at Grandin with a little displeasure. Then he grinned, swatting Stan warmly on the shoulder with his free hand. "Don't worry Dr. McPherson, we still have secrets that we keep from you blokes on the other side of the pond; and we aren't friendly with the Zom like you Yanks are."

"That is correct," Grandin cut in, "which is one of the reasons why we chose to hide your ship here. The U.S. government does not watch Scotland very closely and the British do not like the Zom. They think your government has been very foolish and they have even been…"

"Hush, Grandin," Standish hissed through clenched teeth, "you talk too much." Then he smiled again, "These people aren't cleared for most of the secrets I've shared with you." The wrinkles around his eyes displayed a fun loving character, but Stan wondered. "In fact, knowing this bloke's history with concealed firearms, I probably should have had him searched before allowing him aboard my station!"

"Your station?" Grandin said with mock surprise.

"Yes *my* station. I'm an official member of the staff here. I'm in charge of security—and welcoming dangerous gun slingers." He laughed, facing Stan and slapping his hip as if drawing a pistol, old west style. "This bloke is now considered armed and dangerous, I added that note to his dossier myself," then he threw an arm around Stan's shoulders and squeezed. He was a very strong man.

"Well, that wasn't an event that I am especially proud of," said Stan, "I'm glad he lived."

"Rubbish, old boy!" Jack grew serious as he steered Stan to one side of the dome; the others followed. "Killing may not come easily for you, Doctor, but for that devil Clell, killing comes as easily as strolling across the park for a slushy. Trust me, from what I know of him you saved your own life that day, and by putting a bullet in 'im, you earned my respect." He leaned closer, "I've been trying to do that for years."

He led them to a place near the dome wall, where a large, transparent tube protruded vertically out of the top of one of the cargo ships, through what appeared to be a special kind of docking port designed for the purpose. Inside the tube stood an elevator car and as Standish turned a handle on the tube, a door appeared. The doors to the tube and to the elevator box inside came open together.

"Into the lift everyone," he said. It looked a bit small for four people, but Standish stuffed them all inside and squeezed in backward behind them. "I've been assigned to show you three around the station, but since you arrived late, you'd better get a good meal in you and some sleep first." He closed the door and punched a button on the wall. "There's a planning meeting tomorrow morning in the base conference room on Level 4, Section 7. After that, I'll show you around a bit. Until then, you rest and refresh yourselves while the station personnel and I start looking your ship over. Ah! Here's your level."

They all got out of the elevator and Jack walked them part way around the circumference of the station. Even with wide walls removed between them, the corridors within the ships were tight and cozy. It felt a little bit like Stan imagined the inside of a submarine would be. Busy aliens of various races moved here and there.

"Visiting Earthling quarters are right through this door; there's a meal already prepared for you there. The conference room I spoke of earlier is one deck above you, and the mess hall is one deck below. Dinner is at seven o'clock—I'm lookin' forward to see'n you there and getting to know you two proper. Please don't wander anywhere on the station until I've had a chance to show you what's where; after all, this is a an intelligence gathering facility, and there are areas that are restricted to all but specific personnel, and require an escort. Enjoy. Follow me, Grandin, I'll show you where your people are."

The rooms lacked any windows, but in every other sense were very much like small hotel rooms. There were separate doors, labeled with each of their names and each of the rooms contained a bed, a desk, a small table, and a comfortable chair. The drawers and closets of each room were well stocked with Crenum style uniforms and other clothing of various sizes; all clean, neat, and pressed. The suite had a common area with a wide, curtained window overlooking the glass center of the tower. There was also a table full of cold Earth food consisting of ham sandwiches, salads, cake, beverages, and other common American favorites. There were computer consoles in both the rooms and the common area, with access to a variety of resources, both Crenum and Telesian.

Stan ate well and showered, changed into a comfortable jumper, and went above to *The McPherson* to gather some personal and business items from his quarters aboard ship before retiring to his room.

The burden of fear and doubt that he had carried for the future of his project was, for the time being, gone. He felt tired yet energized, relaxed yet exuberant, fulfilled yet unsatisfied. Yet in spite of all his conflicting thoughts and emotions, the bed felt good against his back. For the first time in many years he felt he could rest without worry, and the moment he closed his eyes, he slept; dreaming of far away places, alien faces, and domed, underwater bases.

Chapter 11

CIA director Bill Jones walked into the Oval Office where President Atequipa waited, talking with another man. The President stood and offered his hand.

"Ah, Director Jones...you're here. Please allow me to introduce Mr. Garrack; he is...an ambassador."

The director shook hands with the guest, "Ambassador...?"

"Yes, he is alien to Earth, he's a Zom."

"Ah, yes!" He shook Garrack's hand again. "I've been looking forward to meeting a Zom."

"You know of us?"

"Why of course, I was briefed several years ago about your arrival here, and of our treaty with you."

"Well then, as you know," said the President, "Mr. Garrack's people signed those agreements with my predecessor; agreements which my administration honors. Mr. Garrack and I have just been discussing some concerns...complaints, actually...that he has, regarding that business in the Sierra Nevada and how it effects the treaty."

"Yes, we talked about that raid. As I recall...a spacecraft escaped, and the facility was empty."

"We found that the craft matches a cancelled NASA design, a proposal made by the Newton Research Center, promoted by Astronaut

Stan McPherson, we cancelled that project as part of our treaty with the Zom. Garrack and I think that McPherson completed the project with private funding, and without complying with applicable tax and oversight laws."

"My office concurs, Sir."

"Mr. Garrack here thinks that the matter could have been resolved a month or so ago, before the craft could even fly, but that the CIA interfered with their investigation."

"Really? Well I'll see what I can do." The director looked back at the guest, "How may I be of service?"

"Well, I understand you have an Agent Rod Mills working in your office?" said the alien rather tersely.

"We do. A good man."

"Well we have reason to believe that he outed one of our agents, risking his life and compromising his mission."

"That would be Agent Clell?"

"Umm, yes…I see that you are familiar with the matter. We were made aware of suspicions that Stan McPherson was in violation of your laws with regard to space propulsion research, laws that are designed to guide your people to a state of full treaty compliance with us. We assigned Agent Clell to observe Mr. McPherson as a part of our monitoring operations.

"To *observe*, I see," Director Jones nodded. He tried to suppress the irony in his voice; he had prepared a little surprise to spring on this Mr. Garrack, and he didn't want to tip his hand just yet.

"We know that Agent Mills has some ties to Stan McPherson," the alien went on. "Soon after Clell made contact with McPherson, your Agent Mills made inquiries about Clell. Soon after that, Mr. McPherson shot Agent Clell, very nearly resulting in his death."

The CIA Director suppressed a chuckle.

The President's tone became grave, "This is no laughing matter, Mr. Jones."

"No, of course not—sorry, Sir," he regained control. "How is your agent doing now?"

"He is fine, fully recovered; about this Mills character?"

"Well, I am very familiar with the facts on this matter, and I'm sure I can clear it up for you quickly. First of all, Agent Mills did not 'out' your agent. We were already investigating McPherson's activities, and saw an alien make contact with him. We were concerned about the possible implications of this on the treaty and investigated your man to find out who he was—to determine if he was one of yours, or some other character."

"That is understandable."

"Our investigation turned up some rather disturbing coincidences involving Mr. Clell, and so we decided to take steps to protect our citizen." The director said flatly. "I told Mills to warn Mr. McPherson that Clell was dangerous, and that he was not on his side. However, Mills did not provide Stan with any details as to who Clell was, or his mission. What Mills did, he did with my knowledge and under my direction, so your beef is not with him but with me. What McPherson did surprises me though, I don't believe that the information we provided would have been enough to prompt a man like him to kill."

"Oh, but he most certainly did attempt to kill Agent Clell,"

"I'm sorry, I do not doubt the accuracy of your information, Sir," said Jones, "I disagree with your apparent conclusions. Tell me, are you sure that McPherson didn't take that action in self defense?"

"Well, I don't see why you are attempting to blame the victim of this assault. It was McPherson who…"

"I for one am very happy to hear that your agent survived his injuries," Jones continued, "but I am more glad that McPherson got the warning in time to take action…action which may have saved his life. You see, while what he was doing was clearly illegal, it was not a capitol crime…at least, not on *this* planet."

"Umm," the President cut in, "Could you elaborate please? First, how did you know that this Mr. Clell was an alien? You can see that Garrack here looks as human as you or I. It's their practice to disguise themselves so as not to alarm the public."

"Clell wasn't in disguise when he contacted McPherson—I have some excellent photos…"

"Hmmm…and why did you believe that McPherson was in danger from this alien Clell?"

Jones pulled a 3x5 card from his suit pocket and handed it to the President. President Atequipa put on his reading glasses and read some of the names on the card aloud.

"Jim Tetraziki, Dr. Paul Peterson, Terri Phillips, Kerry Bowman, Claire Saunders, Dr. Bill Rossini, Mikki Knudson…"

Director Jones riveted his eyes on the alien. "These, Mr. President, are U.S. citizens whom we found to have had contact with this Clell fellow shortly before they seemed to…well…vanish abruptly. We don't know their whereabouts or what happened to them, but several of them have been missing for some time and are presumed deceased."

The President furrowed his brow. He examined the card carefully, then leaned forward and turned the card over, showing it to Garrack. "Do you know any of these people?"

The alien glanced quickly at it and shook his head. "No."

The President leaned back in his chair and looked at the list again, tapping it with the back of his finger. "Well, I've a good head for names, and I do recall a few of these. Looks like a couple of them were involved in one way or another with our space program. Here is a weapons expert I know, she worked for a prominent defense contractor. I met her at a campaign fundraiser back when I was in Congress. This one here is a reporter; I seem to recall he vanished rather suddenly, shortly after Saunders. It was in the news."

He turned toward the CIA director, "Am I correct? Do the other names here fit that pattern?"

"For the most part yes, they do."

They both looked at Garrack.

"Sir, I..." the Director started.

"Thank you for your time, Bill," the President interrupted him. "That'll be all."

"I do have one other thing, Sir—just in this morning." The director handed the President the "classified" envelope he had carried into the office with him. "I think it might pertain to this matter."

The President looked at him for a moment, then took the envelope and opened it, looking through its contents while being careful not to show them to his guest. He stopped and squinted carefully at the photograph, then read the other documents in detail.

"What do you make of this?" he asked the CIA Director, holding up the photo in his hand, the back of it still facing the alien.

"Everything we know is there in my report sir, but some of the data's still being analyzed. We'll need to borrow some weapons experts from the Pentagon."

The President gave the alien a hard look over his spectacles. "Mr. Garrack, what Mr. Jones here is reluctant to say, is that we photographed some weapon detonations over the Atlantic. His office thinks the ordinance used were some kind of anti-gravity propelled, nuclear fusion based, plasma warhead."

The alien stiffened, and the President didn't seem to miss the reaction, Jones could see him grind his teeth together in an effort to control his temper. "Do your people use plasma based weapons?"

"I cannot discuss our weapon technologies with you, Sir," the alien said. "You know that."

The President glared at him. "The event recorded in this photograph happened quite close to the eastern borders of my country; an area which we watch very carefully. We have shipping that goes through that region, Gulf Stream fishing vessels, commercial airline flights, and a very large portion of my country's citizens live near there. What is the range of this ordinance Mr. Garrack?" The President was still showing the alien the back of the photo.

"Again, Sir…respectfully, I cannot discuss…"

"Is this yours, Mr. Garrack? It isn't mine, and I doubt whether any *other* country on this planet would even know what it is. You know, technically such a weapon might be considered a violation of the nuclear weapons treaties that we and other countries recently put into place. Some of those countries have satellites in orbit, taking pictures, like this one, to monitor our compliance with those treaties. They might think this belongs to me. What do I tell them if they ask me what it is?

The alien did not answer.

"First, I'm told that no other alien powers, other than your people, are on this planet. Then, I'm told that there might be some scientists or spies from another alien power hiding in my western mountains, working secretly with one of my citizens, and that if we show up in force we might catch them in the act and maybe get our hands on one of their ships—but the only ship we actually find there appears to have been Earth-built.

Now, I find out that alien nuclear devices have been detonated near one of the largest population centers of my country, presumably in an attempt to kill one of my citizens. You cannot talk about this weapon's technology. Oh, no! That would be bad! Yet someone among you seems perfectly comfortable with *using* this weapon, a *nuclear* weapon that you cannot talk about, just off the coast of my country. Please tell me I'm misunderstanding what has happened here! Please, Mr. Garrack, tell me that I got some part of this wrong! Explain it to me please, Mr. Garrack!"

The alien just sat, looking at his hands, and said nothing.

"What if one of these things had missed whatever it was aimed at and hit a commercial airliner, or a cruise ship?"

The alien remained silent.

The President motioned to Director Jones.

"Mr. Jones here seems to be under the impression that one of your people may be carrying out a policy of abduction and assassination within the borders of my country. I don't recall ever asking you for help enforcing our laws, Mr. Garrack.

"I don't want to hurt your feelings, or upset your government, but do I need to remind you that you and your people are supposed to be a secret? Issues like these breed scandals, investigative reporters, independent counsels and grand jury investigations—very unpleasant things, which tend to snowball and make it impossible to keep secrets! What happens if we can't contain this?" He shook the photo in the alien's face so violently that it rattled.

Without taking his eyes off Garrack, he pushed a button on his desk, "Find Harold and send him in here please." He carefully put the documents, including the card with the names on it, back in the envelope and dropped it on his desk. "Can I keep these for a while, Director Jones?"

"Of course, Sir."

"I hope you don't have any appointments this afternoon, Mr. Garrack, because you, the Secretary of Defense, and I are going to sit down and discuss this. He may have some more questions for you. He'll probably bend over backwards to be rude—diplomacy is not his strong suit, nor his job—so I hope you have a thick skin. Wait out in the lobby, please."

The alien stood and left, closing the door behind him.

The President sighed and put his glasses back on the desk. Then he looked up at the ceiling for a minute. "You know what I just did don't you, Bill?"

"Yes, Sir."

"I should manage my temper better."

"I wasn't going to say anything, it isn't my place."

"It isn't your place to keep secrets from me either."

They sat in silence for a moment.

"You know that I'll provide any information on this matter that you request."

"I want it all, Jones."

"Yes, Sir."

"Every scrap."

"Ok."

"If this secret project of McPherson's becomes a scandal, I'll have yours and Agent Mills' ugly butts hanging on my wall as trophies."

"Sir…"

"Get out of my office, Jones."

"Yes, Sir."

***

Standish explained base operations to Stan, Claire, and to some extent Grandin, who had never been to Loch Ness Base before. Each ship in the outpost, functioning singly or in groups, performed some specific role in the system. There was a mess area, an exercise and recreation area, and numerous types of work areas. There were also two ships which were part of a rotating shipping line, that undocked from the station occasionally to trade places with other ships coming in, transporting people, parts, and materials. Simple components were built or repaired right there, others were brought in from Saturn Base.

The station crew cleared *The McPherson* of all of her cargo almost immediately upon landing, and used the space both inside and outside the ship for a variety of assembly operations. Grav-lifts, parts, construction equipment, displaced cargo, pieces of scaffolding, and other equipment and materials covered the hanger as the aliens went to work in earnest.

Command structure was loose, but well organized. Grandin was in charge of some of the construction operations, having a very high rank in the overall order of things, but he also worked alongside Claire, pulling security duty under Jack Standish. The base commander, a grandmother and a specialist in battle damage, was one of the main people working to repair the drive ring, some of which had be to be completely remanufactured from scratch. Stan had a lengthy conversation with the woman while she worked, and found her love for her people, and for his, to be even more endearing than Grandin's.

Most of Grandin's people worked at the head of various teams, due to their previous experience with Telesian technology, regardless of their individual rank. Those who had been hurt in the fight with the Zom fighter were evacuated to Saturn Base for treatment of their injuries, after which some returned to continue work on *The McPherson*, while others went home.

A number of the station's crew were cultural scientists, and they converged on Stan and Claire, giving Standish a much-desired respite from their nagging and sometimes ill-informed questions. Every now and then, specialists in a variety of biological, psychological, governmental, or other sub-fields would arrive from Saturn Base and spend a day or two following Claire or Stan around, asking them things and poking them with instruments. Apparently, Jack had never tolerated "being a Guinea Pig" as he put it. Stan didn't mind too much, especially after Grandin explained that it might help information about Telesians to spread a little faster.

Sometimes, the Earthlings traveled in disguise to a secret Crenum office in Aberdeen, to use the telephone or access the Internet. They placed orders through their contacts for things which they needed, which were then shipped to any of a dozen or so warehouses around the world and then moved by truck to remote locations, where they would then be picked up late at night by Crenum spacecraft and brought to Loch Ness.

"That was an interesting choice of products," Jack said to Stan as they drove though the countryside one day.

"I think it'll be needed," Stan pointed out, not wanting to discuss the details. The company that produced the archive that he'd requested needed time to fill such a large order. Stan had arranged it through a series of trusted surrogates, and hoped the company would be able to complete it quickly, and without drawing too much attention. Their sales manager had said it was the largest and most comprehensive research project they had ever undertaken, and the most expensive.

"It's a nasty business," Standish said. "Interesting how the greenies are so different from us on the outside, but inside they're just people...with fears and bigotries. These Zom behave much the same way my ancestors did."

"The Spanish were far worse, I think," Claire pointed out. "We'll never recover most of the artifacts on South and Central American cultures. It's a shame, all that history melted down into gold doubloons."

"At least our story will be preserved, whatever happens," Stan said.

"Yep! Better than Claire here," Jack winked. "All she ever shops for are things that go boom in the night!"

"Hey, there are other things on my list!" Claire said, pantomiming the act of pulling a long, thin, sheet of paper out of a non-existent purse. "Here you go right here; Hair pins. Oh! And there," she showed the invisible list to Standish, "tight fitting blue jeans, make-up...oh, and there's the hairspray, right next to the ASH charges!"

"ASH charges," Stan chuckled, "the ultimate teamwork between technology and high explosives."

"Kind of like us?" Claire teased, wrinkling her nose at him.

***

The red snapper remembered warmer waters. It's creators had dropped it off with its brothers into the middle of the Atlantic Ocean, and it traveled east toward it's patrol area off the coast of the country called "Spain". Small fish swam past, but it ignored them. It never grew hungry, and wasn't even capable of eating. In its mouth was a communications device for contacting it's creators, in its insides were sophisticated computer circuitry and a power system, all alien to this world.

Life. What was life? Was it alive? It pondered this sometimes. It had eyes, after a fashion. It could detect sound with a sensor that ran down its sides, the length of its body, the same as any other fish. It couldn't reproduce, but it knew that there were many others like it—proof that, somewhere, it was being duplicated.

Yes, it thought, *I am alive.*

It saw itself occasionally, when it encountered a reflective surface, and it was beautiful. Its bright red color, its elegant shape, it's shining scales, its protruding lower jaw with lovely, exposed teeth.

It had vast memories to help it live in its ocean environment, to protect itself and to aid it in identifying the one thing that it sought, the one thing that meant more to it than anything else it the world.

The *Cladnic.*

It's burning need to find this one thing eclipsed all the normal instincts of it's breed. It would destroy anything that got in its way. All life except its own was meaningless.

Then it heard something—a human voice—its first clue! As it swam toward the sound, it thought about it, separating it from the many other ocean noises. More voices! It searched for a match, processing a list of names and languages for comparison.

No, it decided, *I have no memory of these voices.*

If it could not remember them, they were pointless to pursue. The fish probe watched the boat pass overhead and traced its shape, those parts of it that could be seen from underwater, and searched its memories again.

No, this vessel has no purpose, no value.

Then it heard another sound. Danger! The warning came from a separate system that processed natural and semi-natural hazards, retained from the memories of its biological counterpart. It turned and swam, but the natural fish movement propelling it through the water had too many limits. It had another system, one that used much power; it would have to rest for many hours.

Into the Dark

Escape of the Nomad

Emergency! Its existence was threatened! It had to escape! It activated its emergency power source and slid though the water, pulling itself forward at full speed. Still that terrible net followed.

Then it saw obstacles swirling in front of it, other fish. It tried to avoid collusions, dodging, evading, but there were too many. Swimming around them meant slowing down. It could hear the net coming closer. A natural instinct told it to escape by turning to one side or the other and letting the danger pass, but other fish blocked its path again.

Then it saw the net closing it in on all sides. It could no longer outmaneuver it, could not outrun it. Hundreds of other fish crushed into it from all sides. Unable to escape, unable to move, it shut itself down. It had to regenerate its energy. It had to wait.

The owner of the boat saw the large red fish lying among the flopping, squirming catch of the day and pointed. His brother-in-law, Jose, went over and grabbed it roughly by the gills, lifting it high over his head.

"Pargo!" Jose shouted.

Then Jose jerked violently and fell to the deck, twitching.

He ran to Jose's side and put his ear to the man's chest.

His heart had stopped.

He pounded the man's chest and began heart massage.

"Clear the deck! Get the de-fib kit…hurry!

Somebody pointed and he glanced up, just in time to see the large red fish levitate over the rail, as if carried by an unseen hand, and drop back into the ocean.

***

Agent Clell was sitting at his desk, going through the many clues that cluttered his office. News stories, UFO sightings, bizarre events—this planet was a circus.

Don't these people have anything better to do than make up this drivel?

If only that pilot hadn't been so stupid! He wouldn't have to be sitting here, sorting through these information weeds. The idiot had had the craft in his sights, defenseless, and had let them escape! He should have known they'd try to hide underwater. Now the entire project on Telesia was in jeopardy, an effort that had been in the works for fifteen years, all because one man had built a ship! One man and one ship stood in the way of the most feared military force in known space. Now he had to find the *Cladnic* before it could escape the system. At least he was confident that the ship was still on the planet somewhere. The craft had taken heavy damage in the fight, and wouldn't have been able to leave the atmosphere.

Cladnic, he thought, *what a stupid thing to call a spacecraft. If it really were nothing, then why was it such a flaring nuisance?*

Maybe it never made it to its destination. Perhaps it lay, even now, at the bottom of the ocean. The Zom had probes combing the Atlantic, but the waters were vast. He had to find clues to narrow the search. His eye settled on a photo, a front-page story among the many tabloids that littered his desk. The headline read:

UFO Crashes into Loch Ness

It was another idiotic attempt to attract attention. A copy of the same ridiculous artist drawing of a "Flying Saucer" that Clell had seen a hundred times—only this time someone had overlaid it onto a photo of some famous lake in Scotland.

They probably used one of those souvenir photos, the ones that they print by the hundreds and sell to tourists so they could show others how they were able to travel such a short distance on such a small planet. As if anyone would brag about going to see yet another castle ruin on a world covered with antiquities!

The attached article speculated about the UFO's purpose for being there, that the strange creature that used to inhabit the lake had been a product of a similar crash, and that maybe this craft was delivering another one.

No wonder these people are so backward, thought Clell to himself, *they're so enthralled with fiction they don't care about reality.*

The article contained many quotes from so-called experts and eyewitnesses, all lying; no doubt in exchange for the money that the fringe media paid people for such nonsense.

He was about to set the paper aside, when he noticed something blurry at the edge of the drawing. He looked at it carefully, then quickly set an image of the McPherson craft next to the tabloid. He looked at the edge of the drive ring and then looked again at the bulge that seemed to be poking out from behind the fake spaceship.

They had photographed a real spacecraft! His heart leapt as he looked at the tree lined shore; there were no shadows, and the photo was very grainy. Whoever had taken the image had shot it in dim light; the blue sky in the image was as phony as the silly drawing of the UFO. When they had enhanced the light in the picture, to make the details show up better, the changes further washed out the already blurry spacecraft—so they had placed their own fake UFO over the top of it!

He got up and left his office, the tabloid in his hand.

***

"Have you ever been to Saturn base?"

"Shh, not so loud." Grandin answered, he was heavily made-up to look like an old man from Earth. He and Claire were pretending to be tourists, standing alone on the roof of Aldourie Castle, looking out over Loch Ness.

"Well have you?" She was dressed and made up to look like the old man's wife, and was using the coin-operated telescope mounted on the castle wall to scan the opposite shore. Grandin was doing the same, only he used a camcorder with a telephoto lens.

"Yes, many times, it is much larger than this one. Like a small city. We established it twelve years ago. Many people work there with their families, some have spent their entire careers there. You might call it our home office."

"There are families there? But how? Isn't it dangerous, so close to…you know…" she lowered her voice some more, "the Zom?"

He laughed a little. "People everywhere have the same challenges, in one way or another, space and expediency get in our way too. That facility, like this one, is not very easy to find, not when you consider that both are somewhat out of the way and outside of the Zom's area of interest. There are gas giant planets like Saturn everywhere and no shortage of the resources that such planets provide. We use their gravity for pushing our ships to light speed, and for not much else. They can be very difficult and dangerous to land on and search and anything hiding on the surface of one is just too out of the way for anyone to stumble upon accidentally. Your people have an expression about 'hiding a needle in a haystack'. We know that the Zom are searching the Atlantic Ocean. There is nothing there for them to find, but they can't know that, so they still have to search it; just to be sure."

"And when you search for nothing," Claire put in, "you're never quite sure when you've actually found it."

"Exactly, and the places to search up there," he motioned upwards with his head, "are truly vast, far more than anywhere here, and very difficult to search. The only way they would look for us on Saturn is if they somehow were to figure out that we are there, or as part of a search of the surfaces of all the gas giants."

"The surface? You hide on the planet's surface? How? I always thought you were on one of its moons or something. The pressure…the gravity. I didn't know that people could live in such a place."

Grandin smiled, "We can, and a full search of a planet like that one would take many weeks. The Zom have no interests on Saturn, and no reason to be poking around on the surface except to find us. If they only succeeded in chasing us away, they would accomplish nothing. They have to actually catch us to prove that we are there."

Claire put another coin in the telescope.

"They may not even know that the ship survived," said Grandin, "it was a very near thing. If they suspected that this lake was the place to search, they would probably be patient and search it the same way they are searching the Ocean—the way that you and I are searching for them right now. Any of those rocks over there could be a Zom probe."

"How will I know?"

Grandin lowered his camera and smiled at her, "Just keep watching them; if one of them moves, tell me."

***

"You people just don't understand!"

Rosh Gimlar couldn't decide which was worse; the Zom or his own Governing Committee. After all they had accomplished, the evidence they had gathered, direct charges against the Zom, detailed images, confidential Telesian government documents talking about the Zom, specs on the Telesian spacecraft, recorded interviews with the humans residing at Loch Ness Base—yet still the Crenum government refused to act!

The official in his office, one of the fourteen legislators representing the Crenum home world in the Crenum Governing Committee shook his head, "They're all afraid, Rosh. They're afraid of the Zom military. Our commanders cannot assure them victory against the Zom, and without the actual Telesian ship in hand, your complaints in the UCP are going nowhere. The members of our alliance are scared too, most of them just want us to drop the whole matter and let the Zom have Telesia; after all it is in their space."

"Has the proximity of this Telis star to our border outposts gone completely unnoticed by them? Are they blind? Do they honestly believe that the Zom are interested in that particular planet for biological mining alone?"

"Those issues are huge, Gimlar. You know that. The Zom have more bases on dead planets, and more space to traverse between them, than any other known power. The only thing that limits their growth right now is the ability to feed all their people who are not on living worlds."

"Exactly! The Zom are building an empire, and if we don't stop them here…"

"There are many worlds, Gimlar; if we don't stop them here, we can still put off the danger and try and stop them somewhere else. Don't get me wrong, I agree with you on all of this, but from the things that I'm hearing they're going to continue making excuses, finding reasons to take no risky action. People want to know what we stand to gain from interfering with a world inside Zom space. I'm losing this debate, both on the floor and among the people. Folks *are* upset about the Telesian issue, and they *are* willing to go to war over it, provided they know they we can win that war. Nobody sees that happening here. They don't want to lose a war with the Zom."

"The Zom will turn Telesia into a military outpost!"

"We all know that, Rosh, no one is that stupid. That's exactly why no one wants to anger the Zom. They've given up on Telesia and they're acting as if the Zom already own it. A Zom garrison on Telis 3 is very dangerous for us, but everybody behaves as if it has already happened."

"What about those images that the Kreshans took and released to the media with their complaint? If that's not outside support…"

"It is, but the Kreshans are a scientific people, they are pacifists and have only a small military force. They wouldn't be able to help us in a war against the Zom and that's all anybody cares about right now. The people have heard of so much evil concerning the Zom, they've grown numb to it. It has crossed over from anger into fear. They don't want the Zom to bring that evil home to them. The Zom are very powerful and barbaric and our people are more frightened than angry."

"Might Makes Right," Gimlar said in a low voice.

"What was that?"

"A Telesian expression that someone passed on to me recently, 'Might Makes Right', it means that the person with the strongest arm decides what is right and what is wrong."

"Unfortunately that's the essence of this situation. I've heard a popular Zom expression that fits even better, 'An angry coward is the lowest coward of all'. Gimlar, I'll continue to fight for direct intervention in Telesia, but this effort is about to die unless you can provide me with military support from elsewhere in the UCP. Either that or prove that we are all doomed if we don't act. There are no moral issues here that matter to anyone anymore…and if any more bad news comes out of the UCP or Telesia, then our entire involvement could be terminated in one vote."

"I hadn't thought that things were that bad."

"You have to deal constantly and directly with the Zom. You, Klicktah, the UCP General Advocate, all of those who are working in the Telis system—you are the wise and brave ones in all this…" he shook his head. "…we in the government, along with our people and our friends and counterparts in the UCP…we're all scared blind."

Chapter 12

The eel lay in shadows, in the mud of Loch Ness, listening to the water. It had heard a voice, but in the echoes of the narrow lake, it couldn't quite find where the sound had come from. It had patrolled up and down the full length of the lake for several days, and had heard many voices, only to find them coming from people without value, somewhere on the surface or along the shore. Now it had gone deep, hoping to find the thing that it sought.

The voice that it had heard was too faint to identify, barely loud enough for it to hear.

Then the voice was joined by more voices. Small clouds of silt rose as the eel lifted its head and began to swim, slowly, in ever widening circles, gradually drifting it's search toward the sounds. As the voices grew louder, the probe isolated several and processed them, separating and comparing their words with the information stored in its computer brain.

Identified: Crenum language

It swam faster, following the contour of the bottom as it descended deeper.

After several hours of searching, it bumped into a large building of some kind from which the voices seemed to come.

Recording all that it saw and heard, it swam up, down, and around the structure, learning its shape, until it came up with another match.

Identified: Crenum Environment Module.

It examined a part of the tower closely, to try to find its composition.

Identified: Crenum Exploration Vessel

These people and objects were not what it sought, or had expected to find here, and it could think of no reason for the tower's existence. Even so, these findings were valuable clues. Its program activated a priority 3 action. It obeyed and began swimming to the surface to report.

Then it stopped. There was a thin beam of light shining out from the dome at the top of the tower. The probe moved closer to investigate, sliding close and putting a tiny black eye against the spot where there appeared to be a scratch in the dome's opaque lining.

It saw a spaceship inside the dome, half encased in scaffolding. The craft appeared to have a drive ring, and there were many Crenum busily working on it. The probe traced the image under the scaffolding, found the shape, and then quickly compared it to the template of the vessel that it sought. Immediately, it came up with another match!

Identified: Cladnic – Experimental Telesian Interstellar Spacecraft

It had achieved its purpose! It had found the *Cladnic*!

It could waste no more time. It rose swiftly up, preparing the coded message as it swam.

Once at the surface, it poked its head above the bobbing waves.

The rain that had been falling when it entered the lake had ceased, and several stars shone through breaks in the clouds. It knew that it came from the sky, knew the stars. It remembered some of them, identifying them first by type and magnitude and then by name and coordinates. By these stars, it located an invisible spot in the sky where it knew there was another probe waiting in space to receive its message.

It pointed its long body in that direction, opening its toothy mouth to reveal a small communications dish.

It began to transmit.

***

"You need any help up there?" Standish craned his neck, looking up at the underside of *The McPherson*.

Claire Saunders was sitting on the scaffolding, working on the back end of the weapons platform. A large box hovered on a grav lift next to her and a pile of packing material lay on the floor under it.

"Naw, I've got it, but I'm gonna have a talk with the green guys who built this thing, it's a pain in the neck to load."

"What is?"

"This rotary ASH charge launcher, but I should have designed it myself. I'd have built it to store the charges in a clip, and load it by swapping the empty clip for a new one, but the way these Crenum did it I have to hang up here and insert each charge into the thing one by one, like restuffing a Pez dispenser one Pez at a time."

"Looks like they put it at a funny angle too, easy to reload in a spacewalk, but you need a trained monkey to do it in normal gravity."

"Exactly!"

He laughed, "Maybe we should bring one along."

She climbed down and joined him on the hanger deck, wiping her hands on a rag. Jack noticed that the alien pistol she always wore was still on her hip.

"I hope you don't have to do that to send out those little probe things."

"No, they have docking clamps that they plug themselves into outside the ship. I have to be more careful with ASH charges though, they're not very smart and I don't completely trust them. One good thing, the launcher is EM shielded with a direct uplink port for each charge, so I don't have to worry about someone cracking their code and setting them off. I can also jettison the whole launcher if the things start to misbehave."

"An ASH charge jettison—self-propelled, shaped charges in a can. Nasty surprises, elegantly packaged," Standish said with a grin. "Claire, I was wondering if you could discuss something with me in private."

"Sure, but keep your hands to yourself—I know karate!" She started circling him; shadow boxing. He ducked and laughed.

"Whoa, nay, relax. It has nothing to do with that."

He led her down to the shuttle bay near the bottom of the tower where there was a small, two-seat Crenum submersible. Claire gave him a funny look as he opened the canopy and activated the tiny vessel's systems.

They both opened nearby lockers with their names on them and got out the special life suits that they had to wear whenever they operated this particular craft.

"What's this about?" She asked, as she pulled on her suit and took the helmet that went with it.

"In a minute," he said. "Is your power cell fully charged?"

She checked the indicator, "Yep."

Jack took the front seat and they closed the canopy on the cramped vehicle and launched it out into the water, slipping through the darkness for several minutes. Then Jack shut down thc drive and let the shuttle sink to the bottom.

"Check your scanners, anything nearby?" he asked.

Claire looked at her screen. The vessel had no airlock, but the controls were large and waterproof, so that the occupants could fill the craft with water and operate it in life suits.

"All clear, Captain!"

Jack shut down all power, making sure that the cable that powered the craft was completely unplugged from the power cell. Then he took the energy scanner he'd brought with him and activated it. It was sensitive enough to detect the tiny nerve impulses of a beating heart from as far as 10 meters away.

Claire, looking over his shoulder, started jabbing him hard in the ribs with her fingers, causing pain impulses to light up the scanner's screen.

"Get off!" he said. He held the device up and studied it closely. All it showed was the biological activity of the two humans in the shuttle, and the life support systems and other equipment normal to their life suits. It also showed Clair's weapon that she still wore. Jack's weapon did not show, since he still preferred firearms.

Satisfied, he reconnected the power and pushed a button to flood the cockpit. The shuttle's environment systems sucked out the small amount of air in the cramped space and preserved it in tanks, so that the craft produced few bubbles that someone might notice at the surface of the lake. Once the ship had filled with water, he opened the canopy and climbed out, motioning for Claire to follow.

After they had walked for twenty or so meters, he stopped and activated his life suit's shields, signaling to Claire to do the same. Then he took a small wire from a pocket in his suit and handed her one end of it, inserting his end into a plug under his helmet. She immediately understood and plugged in her end.

"Can you hear me?"

"Ya."

"Have you figured out why we're out here?"

"You used that energy scanner thing to look for hidden listening devices. Activating the life suits' stealth shields blocks all sound from going from us out into the water. So I'm guessing you want to discuss something that you don't even want the fish to hear."

"Good girl."

"So what's this about?"

"How long have you worked with the Crenum?"

She paused.

"About three years."

"I've been with them for five…do you trust them?"

"There is only one person I trust, and it's not you."

He pondered for a moment—had he heard that line in a movie somewhere?

"Again, good girl. I've watched how you look at McPherson, and I think I'm a good judge of body language." He winked when he said this.

"Is that what we're here to discuss? My love life?" Claire said with annoyance. "It'll be a short conversation—he hasn't come around yet."

"The Crenum are acting kind of funny."

"Go on."

"It's this Zom thing, nothing the Crenum do to stop their invasion plans seems to work. They pressure, they persuade, they intimidate…the situation is not good."

Claire looked thoughtful.

Jack continued, "I'll spell it out for you, Claire. The Zom are scary. They're as bad as the worst tyrants in our planet's history. They have all the other greenies worried stiff, the Crenum people, their government, their allies, everybody. Grandin answers to both their military and directly to the Crenum Governing Council. Grandin's military superior, his CO, he's a solid guy, but their government's nothing but a pack of cowards. Word has it they're about to pull the plug on this whole project. Grandin's brave, but he's a man of duty. He cares a lot, but he does what he's told."

"What about you?" Claire asked, stepping closer. "Grandin's a good little soldier; he'll act like a soldier. You're a good little spook; you'll act like a spook."

"Ya…and you're a good little thief."

"Thank you," she squeaked with a smile. She tipped her head playfully and did a little curtsey, stirring a cloud of silt under her feet.

Standish grinned and continued, "I think things are going to go south with the Zom, real bad south, and when they do I don't think we'll be able to rely on Grandin anymore. One day they, his government, might order him to turn on Stan. Glem's a great guy and all, and he won't like doing it, but he'll obey orders. I don't think Stan will take enough steps to protect himself, he owes the Crenum."

"I owe them too."

"So do I, but I'm a spook and you're a thief and spooks and thieves…we don't trust people very well."

"True."

"Do you believe me when I tell you that I'm more loyal to Earth than I am to England? More than anything else?"

"Ya, but I don't always know where you're gonna with that. You know? Sometimes you spooks justify things that make normal people squint their eyes and say, 'Huh?'"

"I firmly believe that everything I do to help Stan helps Earth."

She remained silent.

"I know, spooks lie," he laughed, "and thieves steal. Grandin trusts both of us way too much, but I look in your eyes and I firmly believe that the only person or thing in this universe that can safely trust you is Stan McPherson. Take care of him. Watch his back. Keep your eyes open and be prepared. Whether or not you let me help you with those preparations is up to you, but be ready to protect Stan and his ship."

"Gotcha. Can we go now? My suit's starting to chafe."

Jack laughed.

"Ya."

***

"We have to leave!" Grandin grabbed Stan's arm and pulled him off his workbench.

"Wait," Stan said, and reached back to put his tools away.

"Now!" Grandin commanded as he half-dragged Stan through the door and into the corridor beyond.

"Where are we going?" asked Stan, confused. "What's wrong?" Grandin's urgency worried him, the alien commander had seemed relaxed for so long.

Into the Dark

Escape of the Nomad

"We have to leave the station—the Zom have found us," said Grandin. "Hurry!"

Stan needed no further urging and ran beside Grandin toward the elevator tube. As they went, they had to dodge around station staff as word spread and everybody scrambled for their ships and started preparing to break up the base and turn it back into a fleet of interstellar spacecraft. Stan could hear the hiss of airlocks behind him—the crews seemed to be waiting for them to pass before sealing up their ships for departure.

The ride in the elevator tube seemed to take forever. They could see pilots activating flight systems in cockpits long silent, and technicians standing by with cutting tools in hand, ready to quickly and violently disassemble the elevator and free their ships for escape.

They both hit the hanger level at a run and found the ship already cleared of scaffolding and the equipment stowed. Pumps were starting to flood the hanger deck, and Claire had just arrived at the bottom of the ramp from *The McPherson*, pushing a grav-lift with a heavy crate on it.

"What's this for?" Grandin slid to a stop.

"Zom ships have surrounded the base. Your people tell me there's no time to send out shuttles to disassemble the dome, so I'm gonna blow it with ASH charges. Standish is aboard right now programming them."

Stan and Grandin helped her pry the lid off the crate with a crowbar. Inside lay the explosives, each packed carefully in its own bubble wrapped section of the sub-divided box.

Claire took out one of them, pulled off the packing material, and pushed a button to activate it. Then she tossed the charge carelessly into the air. It rose under its own power, speeding across the hanger to the dome. It then turned and fastened itself to its appointed location along one of the seams between the dome sections. Grandin and Stan looked at each other a moment, and then quickly followed suit. The three of them unpacked the entire crate, unwrapping, activating and tossing each charge.

By the time they had finished, the water on the hanger floor had already risen above their ankles.

"Ok, that's done, let's go," Claire said into her com as they threw the crate off the grav-lift and pushed it back up the ramp into the ship.

On the bridge, Standish greeted them from his computer console. Claire, Grandin, and Stan each took their seats.

Stan took the controls and looked out, "Pumping water into the dome is too slow; can we blow the top charge?"

Grandin nodded and activated the ship's shields as Claire pulled up the controls for the ASH charges on her attack computer. Selecting the charge at the apex of the dome, she punched a key to detonate it. There was a loud "POP" outside and water cascaded in on them until they found themselves once again in the murky waters of Loch Ness.

"Some of the other charges were knocked loose by the blast and the flood; I need a few seconds to get them repositioned," said Jack, working at his console.

"Grandin, are your people in the station ready?"

"They've been ready; they're just waiting for the dome and us to get off of them."

"Well then let's get going. Jack, ready?"

"Yep."

Claire?"

"Right-O." She pushed a button on her screen and they heard a low rumble as the dome blew open, the shattered debris blossoming outward like a budding flower and fell to the lake floor.

They all looked at their scanner screens and the view was heart stopping. No fewer than thirty active sonar sources surrounded them.

"They can't see us can they, Grandin?"

"Only if we move, our stealth systems nullify all of the sonar and other sensors that can see the ship directly, but they can detect the turbulence caused by motion."

"Why haven't they attacked yet?"

"I don't know. I think they were trying to negotiate the surrender of the base when we blew the dome and surprised them. Perhaps they did not think we would try to get out that way."

Stan chose a blip and approached it, gaining speed. As he did, several other blips started toward them. They could see the Crenum base on the scanner behind them, lit up by the Zom's active sonar like a lighthouse on a hill.

Many more than thirty enemy contacts were soon visible. Some were following *The McPherson* while others advanced on the base. The base tower started to dissolve as the Crenum ships undocked and activated their stealth systems.

The blip that Stan was approaching began moving to one side, apparently to avoid a collision. Stan changed course to follow it. Then, the blip disappeared.

"They shut off their active sonar," Grandin explained, "watch out for those behind us. If they are attack shuttles, they will try to dock and board."

Stan spun the ship around suddenly, diving to just over the lakebed and heading back toward the base, which was now half the size as before. He began swerving the ship from side to side, obscuring the sensors with a rising shroud of swirling peat and silt.

"So much for motion sensors," said Claire. "That's why the Zom were staying off the bottom, they didn't want to disturb the mud."

"Claire, Jack," Stan suddenly had an idea, "start launching ASH charges. Send them to the bottom and prepare to detonate."

Stan swept *The McPherson* left and right, in wide swinging arcs across the lake floor, churning up more silt and leaving behind ASH charges like a trail of breadcrumbs. The explosives burrowed themselves into the newly loosened mud.

Grandin spoke up, "All ships have un-docked from the tower, Loch Ness Base is no more."

"Claire, detonate the charges."

With a squeal of glee, Claire punched the button.

A storm of swirling bubbles, accompanied by erupting lake floor, rose up and swallowed them.

"That should give enough cover for an escape," Standish said with a grin.

"We're the ones that the Zom are after," said Grandin. "The further we are from all these Zom and Crenum ships, the safer everyone will be."

"Claire," said Stan, "what do you think? Is it a good day for space travel?"

"I don't know—what's your sign?" she said, pulling up the navigation computer.

Stan laughed, "Capricorn."

"Well, Stan dear," said Claire, "The planets indeed smile on you today. Uranus looks down from almost directly overhead. The half-moon is still out in the South-West sky, with Jupiter in conjunction. Neptune, the god of the sea, is just now setting in the western horizon."

"And Saturn?"

"Way off to the East, toward the Sun."

"Let me have a go at the interpretation," said Standish.

"Go ahead, Jack."

"Neptune, setting in the West, is telling us that our days of protection by him are over."

"Right!" said Stan.

"Saturn, living with the Sun, tells us that we have a bright future in that direction."

"Right again," said Claire. "Hey you're good at this game."

"Uranus, way out in the open like that, is warning us not to go straight up…too obvious."

"Ok," said Stan.

"And Jupiter is calling to us from behind the moon, as if hiding, saying that he is our path to safe escape."

"My thinking exactly," said Stan, and pointed the nose of the ship straight at Jupiter. The Zom would think they intended to leave the system, via the nearest, deepest gravity well available.

"It's broad daylight outside," Claire grinned. "The tourists at the castle ruin are in for a real bonus."

Stan turned to Grandin, "I wonder if the Zom want us so bad they're willing to be seen leaving the water."

"They won't be, they'll have to wait for dark, and our ships can hide in the mud for days."

"Next stop, Saturn!" announced Stan, and for the first time since the ship was built, he shoved the speed controls all the way forward.

Out of the depths boiled *The McPherson*, reaching Mach 1 while still underwater, and exploding from the lake like a missile. No longer a creature of the deep, born of Earth, built in darkness, it left the ancient lake of dark legends and streaked across the sky in the blinding sunlight, dragging a hissing column of water behind it.

Adrenaline coursed through his veins. Astronaut Stan McPherson, after being earthbound for fourteen long years, was finally going home.

What he'd remembered as a thundering, twenty-five second, curved flight to orbit, with the roar of rocket engines in his ears, was now a quiet, straight, eight-second lunge. The eyestrain of emerging from the dark lake into the sunshine was painful, but he refused to blink, refused to miss the change from blue, to deep violet, to black. When the dark came, his eyes had been dazzled as if by a camera flash, so that he couldn't see the stars at first; but the sun and the moon shone through, lighting the bridge from their appointed posts in the heavens—just as he remembered them.

He watched as a satellite, one hundred miles away, zipped past them like a meteor. No space launch had been as much fun as his first; but this trip spoiled them all, even without the G-forces shifting his internal organs and the kick in the pants of rocket engines under him, the astonishing speed with which he now flew exhilarated him. He'd looked forward to this trip for over a decade, tried to prepare himself for it, but he had never expected this long awaited next ride to space would come at him faster than his mind could adjust.

When he finally came to himself, he noticed that his mouth was hanging open, and he shut it. He had left all Earth born powers as far behind him as the slender-winged gull leaves the crabs on the seashore.

He looked around and saw Claire Saunders and Jack Standish still gazing, open mouthed, just as he had a few moments before. He looked at Grandin, who was watching him with a wide grin on his face. Grandin took a breath to speak, but Stan put a finger to his lips. He looked at his watch; just twelve minutes had passed since he had been sitting at that workbench on the base. His heart stopped pounding in his ears and he relaxed.

Finally, Claire came out of her trance and remembered her duty. She looked down at her scanner screen.

"Umm..." she stammered, "There...there is no sign of pursuit."

Standish blinked, "By that do you mean you can see no one chasing us, or you can see no one moving as fast as us?"

"Uh..." she paused, trying to work out what he had just asked her, "Uh, both...I mean neither. It just doesn't seem like anything alive can move this fast does it?" She looked at her screen again, "even though we know that the Zom and the Crenum do it all the time."

"And now you have done it," said Grandin, in an official sounding voice. "The three of you are no longer grounded; you have left your planet, aboard an interstellar spacecraft of your people's own design—that is your right of passage. Welcome to the ageless darkness, fellow Spacelings."

Five minutes after launch, they passed the moon like it was a milepost on the freeway.

***

Gimlar was having a very good day.

He liked receiving good news and bad news in the same package, whenever the bad news was nullified like this. He had just learned that the secret Crenum base on Telis 3 had been discovered by the Zom and attacked. At the same time, he learned that *The McPherson* had led a courageous escape that had allowed every one of the Crenum ships to get away, and was now on its way to the larger Crenum base on Telis 6.

Ambassador Fairdock was coming to his office with "urgent matters to discuss". Gimlar smiled as he poured himself a drink and worked out how he'd respond. He had something in his desk that he'd wanted to poke at Fairdock, perhaps now was the time. He opened his drawer and pulled out an ASH charge that Colonel Grandin's people had sent him, turning it over in his hands. Most advancing cultures eventually developed similar devices, this one was yet another example of how the Telesians were using anti-gravity in worthwhile inventions that changed their lives—Grandin had had this one rendered inert to make it safe for transporting. It was equipped with a computer, a sophisticated Electro-Magnetic communications system, and anti-gravity propulsion. It could communicate both with a central computer as well as network with its fellow explosives—to function in concert with them. A simple enough thing, but it was advancement—a product that used anti-gravity. Further proof that the Telesians were not as stagnant as the Zom were trying to lead everyone to believe.

These devices, and others like them, were in common use throughout the planet, and the Telesians often deployed a great many of them at once for complex demolitions projects. There was another thing Gimlar found interesting; apparently, the Telesians had designed these explosives to function underwater, in their atmosphere, or in the cold vacuum of space, and even to receive signals from orbiting satellites—even though the devices themselves were not equipped for high altitude flight.

There's an important lesson in that for someone, he thought to himself.

He was sure the Zom knew of these devices—they had been in use for some time. The United States military even stocked similar explosives—designed to be delivered to a target from inside a cluster bomb, and placed and triggered with the help of an orbiting satellite—yet the Zom had never mentioned them in their reports to the UCP.

He was putting it away again, when the door opened.

"Ambassador Fairdock," he stood to greet the Zom ambassador, "I am very pleased to see you. I hope you are well."

Fairdock entered and bowed twice quickly as Gimlar returned the gesture. He noticed that Fairdock's face was already the wrong color.

"Director, Sir, I come with desperately urgent news and demands from my home world with regard to Telesia."

"I see. Well, please have a seat and proceed, Ambassador. I hope there's not a problem?"

"Well, I'm afraid there is quite a large problem. We discovered a Crenum base at the bottom of a deep Lake on Telis 3."

"I am familiar with our science station there, please go on."

Fairdock stopped, "You knew of this outrage?"

Gimlar smiled, "Outrage? Come, come Fairdock. Of course I did. I don't follow all of our cultural science projects, but this one was arranged under my direction, in order to investigate…."

"I am sorry, Mr. Director," Fairdock grew bolder, "I do not share your casual attitude about this. I thought we agreed we would work together to stop interference with the Telesians, now I find out that the Crenum have been operating a base there all along, and you knew of it?"

"You speak as if it was a military outpost of some sort—the truth is we only recently built that base. I told you that I would look into the alien interference on Telis 3, did I not?"

Fairdock paused, "Yes, you did, but we did not mean for you to trespass…"

"Trespass? You agreed that such an investigation was a good idea. In fact, I seem to recall that you insisted that the project begin right away."

"Well, I..."

"You certainly didn't think that we could conduct such an inquiry without feet on the ground in Telesia...did you?"

Fairdock stopped talking. His jaw just opened and closed a couple of times, silently.

"I regret to report that it hasn't been able to find very much yet," Gimlar continued, leaning back casually in his chair, "but we were able to make a couple of unrelated discoveries. I seem to recall that your Ambassador Trelleg over at the UCP spoke of having an interest in any Telesian technology that might be uncovered."

"Yes..." Fairdock said hesitantly.

"Well, then," he became as excited as a young child on his birthday, and he let it show. It would add to the drama. "I'm most happy to report to you that our cultural scientists working out of that base have learned of two rather interesting Telesian advances. One is a weapon, common to both the Zom and the Crenum, which the Telesians seem to be testing; the other is a rather interesting demolitions device that I was able to secure a prototype of."

He watched Fairdock, who suddenly seemed strained. Gimlar opened a drawer in his desk and took his time digging through it. Finally, he stopped and appeared confused.

"I was certain I had it—wait a minute."

He closed that drawer and opened one on the other side of his desk.

"Ah! There it is!"

He produced the strangely shaped device and the stack of documents that went with it, and placed it all on the table. Fairdock squinted at it, confused, as if he had expected something else.

"Ah, I see you haven't seen one of these before. What it is, is a shaped chemical explosive, but the interesting part about it is that it is self-propelled and uses anti-gravity for its propulsion; can you believe it? Here, let me show you how it works..."

"Oh, we are familiar with these devices already, Mr. Director," Fairdock interrupted him, seeming relieved beyond measure. "They were invented on Telesia some time ago, four of their cycles as I recall," he was smiling now. "They're called ASH charges, that one in your hand is manufactured in an area known as Texas, in the United States. The company name is printed there on the side of it."

Gimlar tried his best to look dejected. "Oh, then it's not as new as we thought."

"No, it isn't."

"Well, I'm sorry. When we didn't see it on Trelleg's report to the UCP, we thought—"

Fairdock grinned, "You said you had something else?"

"Well...yes, but since it involves weapons technology, you're probably already familiar with it as well. I'm afraid I don't have a sample of it and it was not discovered by us directly, but by the Kreshans."

Fairdock's eyes went wide.

Good, Gimlar thought, *he has the news already. This'll be fun.* He activated the image viewer on his desk.

"This is a vid of a test firing of what appears to be plasma missiles."

Fairdock blinked, "Really?"

"Yes," Gimlar said in his best show-and-tell tone. "The test appears to have been successful; the target seems to have been destroyed. See?"

Fairdock gazed at the moving image with what appeared to be mixed interest. Then he stopped the playback.

"I have seen this; the Kreshans are issuing a formal complaint with the UCP concerning it."

"Are they? Well that's pretty silly of them."

"The Kreshans are suspicious, as they always are. Can you imagine?" Fairdock laughed. "They actually think those are Zom missiles!"

Gimlar looked incredulous, "Do they now? What do your people think?"

"We are…not sure," he seemed to be studying Gimlar's face for clues.

Gimlar felt a little bit guilty for what was coming next, but he took such perverse pleasure in tormenting Fairdock.

"Well, I heard that the Kreshans were annoyed by the finding, but to launch a formal complaint against the Zom over this? Really!" He said, melodramatically.

Fairdock looked at him, "Well, they seem convinced."

Gimlar seemed ready to end the meeting on that note, "Well, anyway…it is quite fascinating that this primitive culture seems to have come so far, is it not? Our science station on Telesia, besides being a marvelous achievement both in Crenum/Zom cooperation and in benign observation, may yet yield useful findings in the future. I look forward to new revelations from it in our joint interference investigation, as well as new data on the Telesian culture and their technological advances, which seem to be on the upturn!"

Fairdock squirmed, "Indeed."

"What's the matter?" Gimlar asked.

"Haven't you heard?"

"Heard what?"

Fairdock sighed. "Well, I have some difficult news concerning your base on Telis 3."

Gimlar now put on his best "shocked and confused" look.

"Oh, no. What has happened?"

"Well, it seems that our military found it and..."

"And?" Gimlar raised his eyebrows.

"Well, they approached it."

"Approached it?"

"Yes, they made sonar contact with it and, well, it dissembled itself and disappeared."

Gimlar sat back and gave a sigh of relief, "Good, that's standard procedure whenever they're discovered." Then he leaned forward again, "Your people did not...attack them, did they."

"Well..."

Gimlar was on his feet now, his face angry.

"You mean after you gave us your permission..."

"Well, we did not really do that; I mean we did not really give you permission to..."

Gimlar poked the desk with his finger, "You sat right here in this office and said..."

"I know what I said, but we also warned that we would fire on any alien ships we found..."

"But we were working together!" Gimlar tried to look hurt, but inside he was almost bursting with mirth.

Ambassador Fairdock fired back, "Working together? You act shocked at the Kreshan complaint, but your people filed a complaint too, and it has upset quite a few people back home; and besides, we never intended to give the Crenum permission to trespass…"

"Trespass? There's that word again! These are tense times I'll admit, but I thought we had a better understanding in here than they have out there!" Gimlar sat back and ran his fingers through his hair.

"Well, maybe if you'd told me about the facility, I could have prevented this," Fairdock pointed out.

Gimlar sighed, "Looking back, that might have been best, I'll admit. We weren't even sure if that particular lake was the best location. It is a completely mobile base, as you found out, and we were considering moving it to a lake in the United States, closer to where the alleged interference is supposed to be occurring. Was anybody hurt in the attack?"

"We are not sure. One of your ships acted strangely and bluff charged us, and then fired some weapons of some kind into the lake bottom. They were harmless, but the resulting turbulence obscured everything."

"Well, that will be looked into," Gimlar was activating his computer console, trying to appear worried and biting the inside of his mouth to fight back the smile. "I'm afraid that I shall have to end our meeting on that note, Ambassador, I have to obtain more specific news on this matter."

Fairdock stood. "Perhaps it will be best not to reestablish that base again until things cool down."

"Agreed."

After the Ambassador left, Gimlar allowed himself to giggle. He really did enjoy these little meetings with Fairdock.

Chapter 13

August 25th, 2035—Urquhart Castle ruins, Scotland —

"The water over there just started to bubble. People were pointing and shouting and I thought the monster was back. So, I said to my son Bob here, I said 'Bob, point your camera over there and get ready to take a picture.' Well, he just stood there, so I took the camera away from him and pointed it right over there, just as this *huge* flying saucer suddenly just *burst* out of the water!" The man was waving his arms, "I can't believe I got the picture! There it is!" He held up the back of his camera and hit replay to show a moving column of water shooting from the bottom to the top of the tiny view frame.

"We have that footage and will show it now," said the news anchor. The image was full screen, replayed in slow motion, zoomed, and enhanced, a clear video of the *Cladnic* leaving the lake.

Clell sighed and shook his head.

"There you have it ladies and gentlemen; you have just witnessed a craft of some kind launching into the sky from Loch Ness. We will watch the situation and release more as it becomes available..."

It had hit the local news first. Not just those who were visiting Urquhart Castle had witnessed the launch, but people all over the countryside saw the strange trail of water vapor angling into the sky. Even some in the city of Inverness to the north said they saw it.

Once people stopped perceiving it as a pathetic stunt to re-enliven the tourist industry in Scotland, the news had begun to spread globally. The fact that the craft was a UFO, and the notion that it might have something to do with the late Loch Ness Monster, was the driving force among the tabloids, initially, but the front-line news took a different tack.

Over several days, local, national, and international news networks had began shelving other topics to make time and space for coverage of the event, as the issue grew legs and gained momentum. News networks scrambled to buy-up the best available footage and to fly reporters to Scotland, then they preempted regular programming to issue updates to the story. Online news outlets were giving the launch, and the issues surrounding it, front-page coverage, updating the images and headlines every several minutes.

The good news, if there was any, was that the craft was so fast that very few had managed to get footage of the ship itself, but the body of eyewitnesses was far too large to suppress by the usual means. There were also many photos, videos, and eyewitness accounts of the water column and the vapor trail, from many different angles and distances.

The networks had called in government officials and various other experts for interviews related to the sighting. Most of those who had not yet seen the videos started out by saying it was probably just a hoax, everyone else avoided comment. After the mass of evidence made the accounts generally irrefutable, serious experts started to analyze available imagery and offer opinions. Then some "unnamed but credible sources" began to come forward and speak to darkened cameras with electronically altered voices, saying that they had seen this same ship, close up, in the

U.S., in the Sierra Nevada Mountains. Others recognized the shape as being similar to a design once worked on at NASA.

The spaceflight debate in Telesia had begun anew.

"We told those idiots exactly where they could find that base, where the *Cladnic* was at the base, and exactly how to approach it and deal with it," said Clell. "They completely ignored us. All they needed was one shuttle with a demolitions team to sneak in there and blow off the top layer of ships. The explosion and flood would have destroyed the craft and produced more than enough Crenum debris and remains to prove Crenum interference with Telesia. Now we're back where we started."

"No, Clell," said Governor Thrigh. "We're worse off. The search field has been increased by a factor of several trillion, and if we don't find them before they reach Telis 5…"

"We can't be sure they went there. We're not dealing just with Grandin, I've learned that Saunders is with him too, and Standish—expect duplicity."

"Regardless, we must catch them before they reach any of the gas planets or they'll leave the system," the governor shook his head. "I'll dispatch my ship to Telis 5 to cut them off in case they go there and I'll send fighters and shuttles to search the approaches to the other planets, but a search of that magnitude will take time. It's just too much space to search with so few ships."

Clell thought for a moment. It would all be so much simpler if the Crenum weren't involved—watching everything they did. Perhaps that was the key, the Crenum and their all fired hidden bases. Maybe he'd been going about this all wrong.

"Tell me, Governor, how have you been managing in your search for that second Crenum base?"

"We've isolated several possible regions, based on past sightings of Crenum ships. It's been slow going. About the only time we get a glimpse of them is when they jump in or out of the system."

"Which planet do they use the most for that?"

"Hard to say."

"Let me see your data."

"Ok, where will you be?"

"I'm catching a shuttle out to your ship to coordinate the search from there. If they're spotted heading towards any of the other gas planets, we can jump in front of them and send out probes."

"I'll have the data sent to your quarters on board, but I and my staff are going with you. This is too important for me to stay here."

"Really? I'd have thought you'd want to get this propaganda situation under control."

"The propaganda situation is beyond control. The speed of the *Cladnic* was well in excess of anything any reasonably sane person can still call grounded, even though it's still slow by our standards. The secret's out, with too many images of the craft floating around. Experts in their media have started to recognize it as something they built, and it's impossible to expect that all of this will not make it into the hands of Crenum agents, and from there to the UCP."

Clell sighed. "Can we pressure the U.S. some more?"

The Governor shook his head. "The *Cladnic* resurfaced in a different country, and even if the United States and the British had adequate control over their media to stop this, the British government never does what we tell them, and Garrack says that the U.S. President is turning hostile. No, Clell, suppressing the awareness of this people has never been easy—now I fear it's impossible. We've waited too long and they've begun to awaken. We should have taken over the operation of this planet years ago and not asked the UCP for permission. Now it's time to act decisively before things get any worse."

Clell knew where he was going with this.

"You're going to send a message to Admiral Steriz?"

"I already have. We'll still try to find the *Cladnic*, and further evidence of Crenum involvement—it might simplify things for the diplomats, but it won't change anything for us or for this planet. We're moving forward."

Clell left the room shaking his head. *"Cladnic",* he thought, *why are we still calling it that stupid name?*

***

"Mr. Jones," said President Atequipa as the CIA director joined him in the oval office, "have you seen the news?" He held up a copy of *Newsweek* with an excellent and dramatic photo of McPherson's ship launching out of Loch Ness.

The CIA Director didn't answer. He stepped forward and handed the President a heavily stamped envelope.

"This is just in from our friends across the pond, Sir."

The President took the envelope and opened it. He read the header, and then thumbed quickly through the document, looking up at Jones occasionally over his spectacles.

"I don't understand. Who conducted this investigation?" he said as he waved the envelope. "Who is this Grandin fellow…is he with JIC? I've never heard of him."

"Glem Grandin works for Military Intelligence, Sir, but not for the Brits. He's with an alien power called the Crenum Alliance."

The President's eyebrows went up, "Well, his English is superb."

"Maybe someone helped him with his spelling."

The President grunted, "Forgive me, Bill. My sense of irony on this topic is rather twisted lately."

"I can appreciate that. Turns out Claire Saunders works for him—has for years. She may have had a hand in that report."

"Hmmm..."

He finished scanning through the document, then went back to the first page, and began reading it again, more slowly.

"Where is this Grandin fellow now? Can I speak with him?"

"He's in deep hiding, incommunicado, but I know where he was yesterday. He was with Claire and a certain Jack Standish of MI6...and Stan McPherson..."

"They're all with McPherson?"

"Yes, Sir. They *were* in Northern Scotland...but they've left now," he said pointedly.

The President went back to reading the report.

"So I've heard," he said. "Do you know where they're headed?"

"Does it matter?"

"No—just curious."

"Someone in MI6 knows, but they won't tell us; afraid of leaks."

"Can you blame them?"

"No. I'm told that they're still in the solar system though...for now."

"For now..." the President whispered. He read the report, taking some notes here and there. He pushed a button on his desk

"Harold, come in here please."

A few minutes later, the Secretary of Defense, Harold Withim, entered and sat down.

"Could you please excuse us, Bill? We'll only be a minute."

"Of course, Mr. President."

After the door closed, the President handed the envelope to Withim. The Defense Secretary whistled between his teeth as he read through it.

"Sir..." he shook his head, "who's this from?"

"The British sent it to us, but it's from the Crenum Alliance...a Zom rival in this region of space."

"Oh..." he said. "So we're a pawn in a power struggle then."

"Looks that way."

"Well, you and I have already discussed some of these allegations, but the other ones here are new to me."

The President stood up and walked over to the window.

"Harold...you know that I've always disapproved of the past administration's practice of signing agreements with one hand while violating them with the other."

"Yes."

"I vowed to put an end to that."

"Yes, Sir."

"I ordered you to dismantle all of the X-13s that my predecessor secretly developed."

"Yes you did, Sir."

"You never quite got around to it, did you, Harold?"

Withim looked up from the document and hesitated a moment before answering.

"No, Sir...I didn't," he sighed. "How long have you known?"

"Quite a while; is this a frequent practice of yours, disobeying orders from your Commander-in-Chief?"

Withim looked at the President with a steely gaze. "I took an oath a long time ago, Sir, the day I first put on this uniform, an oath to defend this country. It was a judgment call."

The President turned and glared at him. "If the Zom had found out about them it would have been a disaster. Would you have made a 'judgment call' to use them too...without my authorization?"

"Of course not."

They looked at each other for a moment in silence. Then the President sat back down and leaned over his desk.

"I want you to put them on alert status..." he whispered, "...make preparations to deploy them."

"So you're treating this report as credible? Are these Crenum any more reliable than the Zom?"

"Well, the British seem to know them pretty well...well enough to harbor them and exchange agents with them...and like you said, these documents repeat complaints against the Zom that we're already investigating, and then some. If these Zom were a country here on Earth, we wouldn't like doing business with them like we're doing now—they aren't trustworthy."

"Yes, Sir," Withim thought for a moment. "I'm afraid we might not get all the X-13s operational; they've been in moth balls for almost seven years now. I'm not even completely sure they'll be effective anyway, the information we have on Zom technology is sketchy, but those shields they have..." He shook his head. "We haven't even seen any of their larger ships. We have very little data on which to build a battle plan."

"I understand; work with what you have. We'll try and get you more."

"Yes, Sir."

"I also want you to quietly move the 1st out of the country."

"I was about to recommend that, Sir."

"And update the pilots' carrier training."

"Of course."

"Tell the Secretary of the Navy that all furlongs are terminated. Continue sending boats out on schedule, but once they're out, they stay out. Arrange to re-supply at foreign ports as needed. Order all our boomers to launch immediately and go deep. I think there's a war coming, Harold, a war which we will lose. We may have to pull a Saddam Hussein and prepare for the rematch."

"Saddam Hussein lost his rematch and was executed by his own people."

"Don't remind me."

"Ok."

"It has to be good. The Zom'll try and garner support and alliances with other countries, so we have to make the world think that this turkey won't be plucked without something resembling an effective fight."

"We will, Sir."

"Harold?"

"Yes?"

"I hope we're both still alive in six months, so we can discuss your insubordination in more detail."

"Me too, Sir."

***

After flying toward Jupiter for thirty minutes, Stan left the ecliptic, where planets orbit, and flew straight "south" for six hours before setting course for Saturn. It would take them a little longer to get there, taking such a roundabout route, but the goal was to arrive unseen, however slow the journey.

Stan had seen many nice pictures of Saturn in the past, and looked forward to seeing it close-up. As they flew toward it, it slowly grew from a bright star, to a very bright disk, to a ball of light with its rings very visible and clear.

Grandin warned them during the command meeting on the forth day of their flight, the day they would make planet-fall, "It is important not to stare too long at the planet during final approach, it will captivate you and you might forget to mind your instruments."

"Grandin," said Stan, "Saturn doesn't hold many surprises for us; we've had good photos of it, from various probes and telescopes, for many years. There have even been 3-D animations and detailed artists renderings. I think we'll be fine."

Grandin smiled, "Nevertheless, I recommend that we do most of the approach countdown items fairly early, and then spend the rest of the morning practicing 'view management' while we rehearse the final approach."

"View management?" said Jack.

"Yes…even pilots and navigators among my people practice how to withstand what we call Gas Giant Hypnosis. In short, the way to prevent it is to count to three whenever you look up at the planet and, on three, do something that draws your eyes away. It is very important, because we will be falling into grasp of the planet's magnificence at the same time that the ship is falling into the grasp of its tidal forces."

"We don't worship planets, Glem," said Claire. "Honestly, I don't see what the big deal is about."

"Trust me, gas giants do things to you, particularly ones of this type. It is too bad that we could not have had a close fly-by of a simpler planet, like Telis 8, to prepare you for some of the effects. They really are spectacular sights. Not only can they temporarily blind you with reflected sunlight, but also there are things about them that fascinate the mind to the point that they tend to mesmerize. This is particularly dangerous, because if you approach a gas giant wrong, its gravity can tear the ship apart. Many have been lost that way. Often, we have found that the pilot or navigator was staring at the planet at a critical moment, and failed to adjust the ship's heading."

"Ok, why not just stuff cotton in our ears," Claire said with a smile.

"What?"

"Like on '*The Odyssey*', where the sailors stuffed cotton in their ears to protect them from the lure of the Sirens' song and keep them from crashing their ship into the rocks. We could do the same thing right? Block off our view of it somehow?"

Grandin looked thoughtful.

"I have never heard of that legend; it is a very interesting parallel. We have tried blackening the view ports of vessels, much like you say, whenever pilots have been inadequately prepared for the effects of Gas Giant Hypnosis, however we cannot do that here. This is still an experimental vessel, with no data for a pre-programmed approach. We need to see where we are going."

They started practicing that morning, while the ringed orb looked about the size of a beach ball. Grandin led them in the exercise.

"Now!" he said with the voice of a drill sergeant as he paced the bridge—they all looked up at the planet. Then he continued, "One! Two! Three! Return to your work!" They all looked away and continued with their duties, even if those duties happened to be reading status reports or reviewing checklists for the approach sequence. They repeated this exercise continuously throughout the morning.

In his earlier space flight experience, Stan had experienced moments of awe similar to the phenomenon that Grandin had described, yet he worried far more about Saturn's gravity and atmospheric pressure.

The shields would provide temporary protection from the atmospheric pressure, and they had brought along two backup shield generators, just in case, but there was uncertainty about how much stress the equipment could handle and of the Crenum crew's ability to switch the systems over quickly enough in the event of a failure.

Claire's job would be to watch the Crenum instruments that monitored the gravitational stress on the ship. Jack's responsibility was to help coordinate Claire and Stan's efforts, as well as to monitor the propulsion systems, keeping everything properly balanced for a gentle, gradually slowing decent. The Crenum had provided him with extra training and well prepared instrumentation, since such things were not normally his field.

As they drew nearer to the planet, Stan began to find it more difficult to look away. The cloud bands in the upper atmosphere, which appeared fuzzy and out of focus on his visual monitors, revealed their details when viewed live—in 3D—through the transparent bridge window. The patterns swirled and mixed, far beyond the resolution of the camera's eye. He could also see that the planet's rings, which had revealed many secrets about themselves over the decades, showed details close-up that no man had ever seen, and were impossible to give justice to through any camera. His mind struggled to grasp the new truths about gravitational resonance that lay hidden in the twisted braids.

He knew he was one of the first Earthlings to view it this way. How would he best describe it, if someone asked?

Into the Dark

Escape of the Nomad

It is the gem of the Solar System, he thought.

To see Saturn through a telescope, it is brilliant, even from a distance. Its rings give it a unique shape and beauty like none other. A favorite of the backyard astronomer, it often steals the show—the star of the star party.

To see it in a photograph taken by a probe, or a powerful telescope, it is magnificent. The colored banding in the atmosphere, though hazy, and in the rings, cause a person to lean closer, to look deeper, yearning to see more.

To see it out the window of a spacecraft on approach is breathtaking. Un-tethered by the limits of imagery, a person becomes dazzled by its brightness and staggered by its grandeur.

I've seen Earth from orbit and it is truly beautiful—and awe inspiring. It consumes the mind with its size, and with its geologic and atmospheric wonders. But Saturn is almost ten times larger than Earth; with thick, complex clouds completely obscuring the surface of the planet and moving barely fast enough for the human eye to perceive. The colors between the bands, pushed by powerful winds, swirl together in circular patterns that never occur again, and never seem to end, as the rapid spin of the planet mixes its atmosphere like a giant blender.

And those rings! The many bands and colors in their plaited arrangements show a depth and iridescence that is impossible to see from a distance or in a photograph. They reflect the sunlight as a rainbow, that shifts and rolls before your eyes, changing at every angle, making you hesitate to blink lest you miss the next phenomenal episode in their never-ending light show. Their glistening particles, arranged and aligned as if by an artist's hand, seem like they're made specifically to entrance you with their sparkling gravitational dance.

All this at the same time that the planet's shear size reaches out and crushes you with a fearful realization of your own smallness. Nothing on Earth can prepare you for it, because the Earth cannot contain anything so in contrast to your size as is Saturn. It looms over you, growing ever larger, like some great doom in a childhood nightmare.

The eyes, strained by its brightness, want to look away; yet they are ensnared in an all enveloping, wondrous trance; leaving the other senses envious of their torture.

Saturn's many moons, unique and glorious in their own right, struggle in vain for your attention, so enraptured are you with the planet.

So much to see, a person can look forever at it..., thought Astronaut Stan McPherson, space flight veteran.

...forever, and never take it all in.

"Stan! Your flight path!" yelled Grandin.

Stan started as if from sleep. Saturn now completely filled the forward view, as if to engulf them. He looked down at his instruments, but the glare from the planet had left him snow-blind. He blinked and squinted, but could only just see the flashing lights and the flickering hue of his computer console in the comparatively dark backdrop of the bridge. Grandin spun Stan's seat out of the way and dove at the controls.

The McPherson had started into a slow spin. It began to shudder, creak, and pop as enormous gravitational hands tried to pull and stretch it like silly putty. The noise grew loader as both Claire and Jack woke from their own trances and began squinting at their instruments. Grandin leaned on the console to steady himself as his shaking hand worked the stick. Then, slowly, the horizon shifted and the noise subsided as the ship straightened out and pointed correctly at the planet once again.

"Never thought it was possible," Stan mumbled, as Grandin stepped away from the console.

"I warned you," he said.

"Claire, did we take any damage?"

"It doesn't look like it, and we're back in the notch now."

"It says here that we're still coming in a little too fast," Standish said slowly, still blinking. "I'll...I'll try to slow us up a bit."

"As gas giants go, this type is one of the more..." Grandin said, staring at the rising cloud tops. "One-Two-Three," he counted aloud, and then turned his back on Saturn and walked away.

They descended carefully into the atmosphere and through the higher-level clouds. When they leveled off their decent, they found themselves in the midst of a global hurricane where the wind whipped them about, and lightning lit the sky continuously as it struck from the never-ending storms above into a glowing blue ocean of liquid methane below.

"We can relax now," said Grandin. "The shields are holding, and we are well beyond the ability of any probe or ship to see us from space."

"Where's the station?" Stan asked.

"Under this shallow sea. We have some distance to travel though. We must hurry. "

"It's very beautiful," said Claire, gazing at their surroundings.

"Deadly beauty," said Grandin. "There are powerful energies at work here; we have to get to the base and under cover of their shields quickly. As good as our shields are, they will not last long against this pressure. The glow that you see in the liquid methane is not sunlight but Saturnlight. The lightning triggers short nuclear reactions in the metallic hydrogen under this ocean. The base is able to draw on that energy and replenish its shields to maintain a constant safe zone. A spacecraft in flight cannot last much longer than half a day before the planet crushes it."

"No wonder you can hide here, nothing can survive long enough to look for you," Stan said.

"Not without a massive effort that we would notice. We have a hidden probe, on the moon that you call Mimas, and on moons of the other gas giants. If they were ever to become interested in searching these outer planets, the probes would warn us."

Stan leveled out *The McPherson* and cruised low through this surreal setting for several hours. When the navigation computer indicated that the base was near, he brought the ship to a halt, hovered there for only a moment, and then slowly dipped the McPherson into the frigid sea.

All on the bridge, except Grandin, drew a collective breath at the site. The liquid methane, which had seemed opaque before, was much more transparent beneath the surface, and only about two hundred meters deep. Hydrogen "mud" bubbled below them like a pot of soup, and glowed like a huge night-light. There were no chasms, rocks, or similar features. The under-Methane terrain appeared flat for as far as the eye could see.

The blue light all around gave the base the appearance of a city made of amethyst. Where Loch Ness base had been a single, small tower hidden in the dark and obscured by silt and peat; this base was a collection of more than a hundred brightly lit towers of varying height and width. Several of them were huge, the walls comprising multiple ships. Some of the structures were so tall that they almost reached the surface. The light from the windows shone through the natural glow around them like a beacon to tired, lonely travelers. Each tower touched at least one of its neighbors, and at the center of the base were six of the larger towers, each five times bigger around than the base at Loch Ness, with a huge collective dome over them.

Stan could see throngs of people through the transparent domes at the tops of the towers who, when they looked up and saw *The McPherson*, pointed, waved and cheered.

"Looks like some kind of festival," said Claire.

"Actually, you're the festival," said Grandin.

The humans all looked at him.

"The reason this base exists is to observe the development of your people. This is a landmark day—the début space flight of a Telesian interstellar craft, and the first voyage of Telesians to an outer planet."

"Wow," said Stan as he looked around at the aliens celebrating, "you folks must like us a lot."

"We're not that happy with Telesians in general, but you three—we like. Put in through that gap in the towers over there."

Once inside the entrance, they waited while the base shields closed them in, and the station's staff pumped the liquid methane from the small space and replaced it with breathable atmosphere. Then they lifted off again and flew through another gap to the center of the tower-circle.

Below them was by far the largest crowd of aliens they had ever seen. Almost a thousand Crenum were crowded together, elbow to elbow, in the open area within the tower-circle, jumping up and down and throwing things in the air. Others leaned out of windows in the towers, throwing what looked like paper.

"It's like a ticker-tape parade," said Jack.

"Land on that X' in the center over there."

Stan squeezed the ship's landing struts into the cordoned area in the middle of the screaming, chanting throng. The instant the ship was on the ground, Grandin, along with several other aliens who had come up from below, grabbed the three Earthlings, McPherson first, and shoved them down through the ship, through the lower airlock and out onto the ramp.

The sound that greeted them was deafening. The paper debris that filled the air flew like a blizzard all around. Upon seeing Stan leave the ship, the massed crowd immediately lost control. They closed in, destroying the barricades and rushing forward as one body; they swarmed the entire crew of *The McPherson*, lifted them onto their shoulders, and paraded them around the courtyard amid exuberant cheering for almost an hour.

***

"They found another one."

"What? Not another floating fish story?"

Phone calls and personal accounts of UFO sightings had swamped the New England Tattler for days. Sweaty, half-baked kooks ran rampant, dragged away from their TV's and out of their basements by the opportunity to cash in on the hysteria. Pie tins, paper plates, hub caps, light fixtures, Oreo cookies, and anything else even remotely resembling flying saucers, were being tossed in front of badly focused cameras in backyards all over the world. Ever since the images and films of the spaceship launch at Loch Ness had hit the news, the staff had been working overtime trying to come up with new and interesting angles that the regular papers had not already used. Now this was the fifth sighting in a week of mysterious flying fish.

"Tell them our rates for photos and..."

"This isn't a photo, it's in a box!"

"In a box? Where?"

"Downstairs in the break room."

It was in a wooden crate that, judging from the smell, could have been at the fish market just yesterday. Today it contained only one fish, but not really a fish. A single black grouper was in the crate with a large, ragged hole through the side of it. Strange looking mechanical and electrical "guts" were visible through the hole, and a tiny satellite dish of some kind hung by wires from the fish's large, gaping mouth.

"It came from a Nantucket fishing boat. They said it electrocuted one of their people and was flying towards the rail when the captain blew it away with a shark gun."

Twenty minutes later, five men in suits loaded the crate into the back of an SUV and drove away.

***

Claire walked into Jack's office in the former Loch Ness Base, now rebuilt on Saturn. After the celebration, they'd moved the ship there, and he'd resumed his duties as chief of security.

"Could you please pick up a shipment for me when it arrives?" Claire said as she dropped a note on Jack's desk. "It's all right there."

He looked at the paper. It was a handwritten note to take a shuttle to a particular location on the fringe of the base, and bring back several crates of equipment that would arrive there in two days.

"What for?"

"I have to get back to Earth."

"But we just got here—homesick?"

She laughed. "Hardly. No…we left Loch Ness in a hurry and there are some things I ordered; I need to pick them up. I know…I was looking forward to seeing the base, but this is important. Keep an eye on our friends for me while I'm gone."

"Ok…take care." He waved the note in the air, "What's this about anyway?"

She glanced over her shoulder as she strode out.

"Ask the fishes."

Chapter 14

"Why do you need four invasion fleets?" asked Admiral Steriz.

"In case I have to respond to Crenum intervention," replied Admiral Taff.

"We've already made plans on how to deal with any Crenum action in Telesia."

"Who drew up those 'plans'?"

"I did."

"I thought so; I had a look at them. They have your arrogance and short-sightedness written all over them."

"You're out of order and insubordinate," Steriz replied calmly over his drink.

"I have a job to do; you will assure me that my requests are granted?"

"Yes, the orders have already been issued. The forces that you requested are in route. Do you understand how expensive this makes things?"

"Do you understand what will happen to those troop transports if Crenum attack cruisers start showing up?"

"Our sources within the Crenum government say that there will be no intervention, the Crenum haven't the stomach for it. Still, I suppose there is a risk they might change their minds. What is *your* plan?"

"My plan is that if the Crenum interfere, I'll use those extra fleets to strike their three border outposts and split their forces."

"We are not at war with the Crenum."

"Don't treat me like I'm stupid, Steriz. War with the Crenum was the goal all along and you know it."

"Yes, our superiors do seem determined to push us into one don't they—even if it puts us at war with the entire UCP. Do not underestimate them, Taff, my gut tells me that the Crenum have some surprises in store for us; that's why I decided to send you in the first place. Now, a separate strike like that will require a separate commander."

"Yes, it will."

"My price for providing you with those extra forces is that I pick who will lead the attack against Telesia."

"Fine, who is it?"

"Admiral Shellar."

Taff snorted. "He's an idiot."

"I know, but he has connections, and my position requires me to feed his ego—not an easy appetite to satisfy. Do you think it will take more than an idiot like Shellar to handle a simple assignment like conquering a grounded world?"

"Of course not, that's why it's the only task you ever give to him. Operations against grounded planets don't require brains."

"I have to have something for the idiots to do while I send my brilliant admirals into the teeth of the enemy."

"I understand."

"There is one flaw in your plan, Taff, Telesia is far away and ordinarily the decision to trigger a war with Crenumar would not be yours. At the first sign of Crenum activity, don't wait for orders from me. Assault the Crenum as you said—immediately. Leave Shellar to defend Telesia. If we can take their border outposts, then Telesia will no longer be closer to them than it is to us. It'll end up isolated behind our lines, so we'll win it eventually anyway, regardless of what the Crenum do to Admiral Shellar in the short term."

"Where is Shellar now?"

"He's on his way here, commanding your invasion fleets. You see, moving forces through friendly space, in times of peace, is another task that doesn't require any brains."

***

Sgt. Brent Hathaway saw the Fiat pull up and stop. The two Italian Carbiniari looked up and saw the vehicle—then ignored it. It was an Italian car, but clearly an American driver—a blonde woman in a U.S. Air Force uniform. The Italian police were there to screen Italian employees who worked at the base, not Americans.

Hathaway stepped up to the vehicle and looked inside. The woman rolled down the window and handed him her ID. He examined it briefly and handed it back.

"I need to get you a visitor's pass, Lt. Schaffer, could you park over there and come inside please?"

"Sure thing."

After she signed the necessary paperwork, he gave the Lieutenant her visitor ID.

"Do you need a map of the base, Maam?"

"No, I know my way around. Thanks."

"Very well."

After returning his salute, she turned and left.

The other guard watched her go.

"She looks familiar," he said. "Has she been here before?"

"I've never seen her here."

"Hmmm, still..."

"Grace Jenkins."

"What?"

"She looks just like that singer, Grace Jenkins."

"Oh, ya! She could be her twin," he shook his head and smiled. "Crazy woman. I heard that new hit she has out."

"Really. What's it called?"

"*You Left Me, So Forget You.*"

"No surprise—that's all she ever sings about."

Lieutenant Jill Schaffer drove away, listening to her car stereo, humming along and bouncing her head to the harsh beat.

You dropped me flat,
But that's just fine.
Now my stuff is mine,
And your stuff is mine.

You took my heart,
You failed the test.
You took my heart,
Now I get the rest.

She stopped at a stop sign and took a quick look at her map. She had lied to the guard; she didn't really know her way around this base, but she couldn't tell him that—it all had to appear perfectly routine.

She'd "visited" many military bases, and worn many disguises. It was all so easily done, she wondered why more people didn't do it. She sometimes ran into folks who used to know her, so she fashioned her disguises to resemble someone famous. That way, folks would catch the "actress or singer so-and-so" resemblance and miss the "Claire Saunders" resemblance.

She pulled up to the aircraft maintenance office, parked the car, and went inside. She stopped the first person she saw.

"Where might I find Lieutenant Meese?"

"Three doors up the hall and to the right."

"Thank you."

The age of the building was astonishing. She wondered, as she walked down the hall, what the U.S. Military's fascination was with these old, World War II era facilities.

They all look the same, she thought.

She stepped in front of the doorway.

"Lieutenant Meese?"

"Yes, Maam. How can I help you?"

"I'm with the Inspector General's office." She handed him a folder.

He looked through the paperwork.

"It says here that you want twenty-four of our IR imaging modules?"

"Yep, right off of your front line ordinance."

He thought about that for a moment.

"For keeps?"

"Well, I have to look them over first, and then I'll send them to a lab down in Langley, where they have a simulator they'll plug into. After that, they'll be reinstalled into live ordinance and sent to Yuma for test firings."

"Wow."

"You can have them back after that" she grinned, "If you still want them."

"No, I reckon not. I heard some congressman got a bee in his bonnet about friendly fire accidents, is that what this is about?"

"Ya."

He thought a little more.

"All right, you have a seat right here. This might take a while; I'll need to get some help pulling the ordinance and removing and packing the modules."

"Nothing doing, soldier," she shook her head. "I'm not just some office flower; I worked in ordinance before I got this job. I'm gonna help ya!"

He looked at her clothes, "You might get that uniform dirty."

"I came prepared."

He grinned, "All right then, let's go."

They pulled examples of several types of missiles, tagged them out, and set them on work racks. Then Lieutenant Meese opened a panel near the front of one of them, removed a small grey box, and handed it to Lieutenant Schaffer—who carefully packaged it for transport. By the time they'd repeated this operation on all twenty-four missiles, it was almost 5:00 pm.

"Good, we're done," said Meese. "My son has a soccer game to-day—it starts in thirty minutes."

"Then I won't keep ya, I can wheel the boxes to the car myself. I'll leave your cart by the door."

He handed her a piece of paper.

"Could you sign this hand-receipt for me please?"

"Sure."

"Interesting overalls, never seen that style before."

She ignored the comment and handed the paper back to him.

He watched her as she packed the last unit.

"I must admit, you do know your stuff."

She tipped her head.

"Thank you."

"Tell me," he said, "we received a shipment of twenty-four of those things just this morning. I never requisitioned them, did you do that?"

"Ya," she said. She turned and faced him. "I didn't want you to be short on inventory. You should be able to program the replacements and have them installed in these bad boys by noon tomorrow."

He scratched his head. "This ordinance has been on the line, unused, for four years. The ones they replaced were on the line for five years; none of those were ever used either. Yet you're worried about my mission readiness. Do you know something about future plans that you're not telling me?"

She looked away. "I just don't wanna shorten your inventory."

"Ok."

"Tell me, Meese," she added. "Do you and your family live on base?"

"Ya, but we're considering moving out into the economy…a nice little farm house in the country I found to rent."

She thought for a moment, then nodded. "Yes, a farm house. That's good for kids. You should go ahead and do that." She looked him hard in the eye. "Waste no time about it."

Then she left, pushing the cart ahead of her.

He just stood there, scratching his head.

***

"Those plasma missiles we photographed—the Zom were shooting at McPherson's ship weren't they?"

"Yes, they were. I was there."

"You gave documents to the British, through your contact with them—-and they passed copies on to us. They said that the Zom were involved in our raid on McPherson's hideout in that abandoned helium reactor facility in the Sierras."

"Yes, the Zom pressured your government to locate and raid our operation there."

"Why?"

"They were planning on you finding us, aliens, on your planet, so they would be justified in coming to your rescue to save you from us."

"But they had fighters searching the area, so that if we failed, they could take you out themselves."

"Yes, and they very nearly succeeded. As it was, a couple of my people were seriously injured in that engagement. It was only clever strategy by McPherson and Saunders that saved us. They used a decoy drone to draw the fighter off so we could escape…I hear that you use that tactic a lot in your submarine warfare."

"We don't talk about submarines. What about the disappearances?"

"I am personally aware of twenty eight successful assassinations; I have a list of names that I can give you. Claire Saunders and Stan McPherson are two that we helped to evade the Zom. There were others, but those were situations where we were able to intervene without revealing ourselves. The British prevented several assassinations in their country as well—they have not told me how many. Their methods were

somewhat more…direct; a number of Section 5 and Zom agents were killed in the process. It is good that you have such loyal friends in the British. They discovered us while interdicting the Zom and we had to pull them into our confidence. They are a very useful ally.

"Why did you reveal yourselves to Saunders?"

"The Zom prefer to take a planet into their confidence…you know, gradually make most of the population into dependent and willing allies, before they take over. They use a lot of propaganda and misinformation to do this, but they know that propaganda usually has an expiration date. If Claire had gone public with the information that she had discovered, they would have found a way to turn it around to mean something else. Then they would have been motivated to take over the planet much sooner—before time had a chance to work the truth out.

"Part of my mission here is to try and prevent an invasion of your planet by the Zom, so I brought Claire onto our team. The arrangement allows her to hide from the Zom and counter Zom activities—as one of us—under my command. I used the information she found as part of my report to my government.

"What about McPherson?"

"(sigh) The Zom took a lot from your country, and your world, in that little agreement they made with you. Not only did it put you in a situation where an invasion of your planet could eventually be made legal, but it also created an environment in which a project like McPherson's could never succeed. He would have been stopped eventually, if not by the Zom then by you. I was under orders for a long time not to interfere, however once the Zom were made aware of McPherson, we knew they would try to get rid of him and his ship. That is when my orders changed and my team moved in. We assisted Stan in completing construction. We tried to limit our encroachments into the ship's tech-

nology as much as we could, but we had to equip it well enough to escape both your reach and that of the Zom ."

"Have they escaped?"

"I'm sorry, I cannot answer that question. If I said more it could endanger the McPherson ship and many Crenum lives. You must understand, Agent Mills, the Zom are deeply infiltrated into your government, even we do not know precisely how far, but everything that we do know is in that report that the British gave you. The people of my country have already risked their safety far too much for you, and the security of The McPherson is now directly linked to our own national security. Our involvement here has left us in an exposed legal situation and the tensions between the Zom and my people have reached a very dangerous level. It is essential that they do not obtain hard evidence of our presence here in your system, or of our contact with you. They could use it as an excuse to engulf this entire region in a very destructive war that I am still trying to prevent. The recording that you are making of this interview must be taken straight to the very top of your government and seen and heard by no one else."

"It will be, Colonel, I assure you. My boss will be taking it to the President, not even our office or the President's advisors will hear it, for now."

"Good."

***

The recording ended—President Atequipa switched it off.

"If they had managed to destroy that ship we would have been doomed, and it would have served us right."

"We might still be."

"Well, I have taken steps."

"Be careful, Sir, you heard what he said, they're everywhere," Jones grinned. "I could be a Zom."

"No, you couldn't—I'd know. You heard that man's voice, there is an arrogance that all these aliens seem to have regarding us Earthlings. I can even hear it in this Colonel Grandin fellow, and he seems to respect us a lot more than the Zom do. It appears that none of these aliens can resist the urge to treat us like children—like ignorant savages. It's a natural reaction, and historically typical, I suppose. They can try to hide it, but smart people can see it if they're looking for it and paying attention."

"Whatever you say, Sir, but we must be careful. I don't even trust my staff with information on this case. Mills and I and a couple of others are all that I have working on it."

"We'll all be careful," the President's voice turned deadly cold, "and we'll be ready. I honestly don't know how to win a war with the Zom, but I can promise you this…they'll come away from the experience a little less cocky."

***

"You're going where?"

"I don't know, were flying out tonight…the pilots, support crews, everybody. They're moving the entire wing somewhere overseas. The pilots'll fly in their planes, everyone else will take commercial flights. They're giving out plane tickets and itineraries at the base and we're leaving from there. We're not allowed to talk about it, and we're leaving right now."

"What does it all mean? Does it have to do with that spaceship thing I saw on the news?"

"I don't know; they're not telling us anything yet. Tell the children I love them. I'll send for all of you...if I can, but if anybody asks, tell them I've gone fishing in Alaska."

He wiped a tear from her eye.

"Come home to me, Captain Forester."

He smiled.

"I will, Mrs. Forester."

***

"Governor, have a look at this."

Clell put a data module in the Governor's computer console and pressed a button.

"What am I looking at?"

"These are the headings of the Crenum ships that have been sighted coming and going from the gas giants in this system," said Clell, pointing. "We've always just assumed they were using those planets to go in or out of light speed, but look at this." He pulled up the sightings around Telis 6.

"It looks like those five were seen going directly to or from the planet itself," said the Governor. "I wonder why we never noticed this earlier."

"Probably because nobody bothered to look. The planet's motion makes it too much trouble. It's a gas giant, so people just assumed the ships were entering or leaving the system—that's what gas giants are for. But while that sort of thing takes place in a high orbit; these vectors show their origin or destination were on a straight line with the planet's surface."

"If we saw them doing that five times, in a lot of random sightings all over the system, then they must use the planet's surface a lot. I wonder why. They couldn't possibly have a base there."

"Maybe there's something special about this gas giant that we don't know about."

"Hmmm...the planet was surveyed many years ago; nothing unusual was found."

"Suppose, instead of leaving the system, the *Cladnic* went to Telis 6."

"Why?"

"To do...whatever it is the Crenum seem to like to do there."

"Ok, we'll stay here at Telis 8 a few more days," said the Governor. "If the *Cladnic* doesn't show up by then it's never going to and we'll jump to Telis 6 and search the planet and moons. If there's something there that interests the Crenum, maybe we'll find out what it is."

***

"Gimlar, I have news that you're not going to like."

The supreme commander of the Crenum Defense Force walked into Rosh Gimlar's office.

"What's your news? Invasion?"

"No. But our probes are in place to watch for one," the commander had a strange look on his face.

"Yes?"

"You know that the military works for the Government."

"Of course, I understand how the relationship between our Government and Military functions, Admiral, is there a problem?"

"Today the Governing Committee went over my head and sent Grandin sealed, irrevocable orders—to be opened in the event that an invasion of Telesia by the Zom becomes imminent and unavoidable."

"A Zom invasion would mean that his mission has failed. Grandin would have to take action to get his people out fast, before any probe broadcast could reach us here, but bypassing you and the normal command structure is most unusual."

"They've done it that way because they know that I disagree with these particular orders. They said they want damage control…they want insurance. They ordered Grandin that if the Zom invade, he is to seize *The McPherson*, by force if necessary, and bring it out here to Crenumar."

"To assure that proof of the reasons for our actions in Telesia is secured, by us, and not lost or tampered with."

"That's the reasoning they used, yes."

Gimlar scowled. "Did the Governing Committee also reason that bringing that evidence here would also render it useless?"

"They care only that it's secured."

"I guess they don't understand the situation very well," he looked at his hands and shook his head, then looked up. "It's my fault I'm afraid. I should have spent more time with them explaining it."

"They're worried that McPherson might try to return to Telis 3 and participate in its defense."

"He might return to Telis 3, Admiral, but he's not stupid. His ship is armed, heavily, with our weapons, but it isn't a proper warship—and McPherson knows it. He wouldn't try to use his ship to fight off an invasion. Besides, he hasn't been treated very well by the authorities of his world; and his loyalties are to that ship more than to his planet or his people."

"Maybe, but he might still feel a need to try to help in some direct way."

"Can you blame him?"

"No, but he can't change the outcome."

"It would be much better for us, Admiral, if he just left Telesia right now and flew that ship straight to some nice, safe, neutral port near at hand, like perhaps in Xentar, or Sisna. Bringing it here would only vindicate Zom claims, triggering a demand that the ship be returned to them in exchange for dropping their plans of retribution against us."

"That might be the real reason for these new orders."

"And Grandin will obey those orders, won't he?"

"Of course he will, he will always obey orders, even if he disagrees with them. That's the way Grandin is—it's the very reason we put him there."

"Well Admiral, no offense to you or your people, but I hope that Stan McPherson is smarter than Glem Grandin, and thwarts him. I want that ship in UCP hands—if not for the Telesians' sake then for ours. On top of everything else, I'm concerned about the effect our actions will have on future relations with the people of Telesia."

"What are they like, these Telesians?"

Gimlar thought for a minute, and then scowled.

"They will make very good friends, but they can become extreamly aggressive and dangerous enemies, and I don't believe the Zom can control these people across so wide a frontier. I've studied the Telesian cultures very closely, and there are many times in their history where a powerful, repressive government like the Zom have been unseated and crushed by seemingly inferior populist movements. I'm convinced that one day the Telesians will throw them off and then either become part of the Crenum Alliance or build their own empire—and that second possibility worries me greatly. If they are our friends, they will enhance our security, but if they become our rivals they will be very difficult neighbors to manage, and may even grow to be worse than the Zom, especially if they come out of this with a hatred of all aliens."

"Well, the Governing Committee knows that the Zom are going to invade Telesia, it's an issue of when, not if…"

"Yes."

"…but you and I know exactly how McPherson and the Telesians who are with him will react if Grandin tries to take that ship away from them."

"Yes, we do," Gimlar sighed, "and we know that people could easily die as a result. That would be tragic and could further erode public support for efforts to help in Telesia."

"Very easily."

"Unless the Zom wait a little longer, until time has taken the Telesian spacecraft out of our hands."

"I don't think they'll wait."

"Neither do I."

***

Under the watchful eye of either Claire or Jack, the aliens carefully removed the old weapons platforms and replaced them. The technicians then mounted strange, new weapons, forty-eight "single-use" plasma missile launchers and thirty civilian-grade particle beam cannons—anti-pirate weapons, typically installed on civilian ships for self-defense. The aliens replaced the Earth-made Aegis defense system with its alien equivalent. They also replaced the improvised ASH charge jettisons with alien probe launchers and installed a new targeting computer—the top-line Crenum military model. Also, avionics, navigation, internal gravity systems, sensors—the aliens upgraded them all.

"We will now work to prepare you for the next stage of your journey," said Grandin. "Your new destination is Redliv 3, in the Sisna Cooperative, over thirty light years from here."

"Let's see," Claire rolled her eyes and laughed, "that's what; one hundred and fifty years travel time?"

"Five months, if we jump off from Saturn," said Stan in an even tone, "much faster if we jump off from Sol."

"Five months, three weeks, four days," Grandin corrected, "and nobody jumps off from a star, it's far too dangerous. Anyway, I wanted to upgrade your propulsion systems to achieve faster speeds but my request was denied. We won't have time anyway, things are growing tense and we have to get you out of the system. Therefore, we have to compensate for your slower speed by getting you ready to leave as soon as possible. The ship will use the planet's gravity to slingshot-accelerate you beyond light speed to another gas giant in the Redliv system. We will remove all construction operations from aboard your ship very soon, to make room for travel essentials—mostly living space and biological supplies needed for the crew during the journey. The trip is very long and time is short." He smiled, "I'm sorry Claire, you won't get to explore this station to the extent that you intended."

"Oh, I got what I wanted," she glanced quickly at Jack, who discreetly avoided her eyes.

"The work outside the ship will be completed shortly, after that it'll be another four or five Earth days to get the ship ready to go."

Stan looked at him, "These 'biological supplies'…"

"Crenum space rations, I'm afraid there will be no time to prepare very much Earth food for the trip, you will have to get used to Crenum cuisine."

"Not like Mom used to make eh?" Standish quipped.

"You know, you've been grumpy lately," Stan pointed out.

"Lack of sleep," said Jack.

Claire looked at him again, and then changed the subject.

"Tell me, whatever happened to that thing Einstein said about nothing going faster than the speed of light?"

"Don't get me started," said Stan.

"Well," said Grandin, "the answer to that is very complicated. Let's just say that his formulas were correct, but actual speed needed to be a zero value constant."

"Huh?"

"What the colonel means," said Stan, "is that nothing goes faster than zero, with respect to itself, ever."

"But, I'm very familiar with the formulas you're saying are meaningless. They say that it really shouldn't be possible to fly from Earth to anywhere faster than light. Where are they flawed?"

"Ok," said Stan. "How fast are we going right now?"

"Ummm…you know that I'm not that familiar with astronomy, Stan. I know that Saturn spins really fast, and we're near the equator…"

"Ya, but none of that really matters does it? Because from the perspective of the here and now, our speed is zero."

"I was thinking more along the lines of our speed from Earth's perspective."

"There you go—that's the mistake folks make. In space, nobody cares what Earth thinks. Our speed, from the perspective of the ground under our feet, is zero. On a craft, flying through space, it's no different.

Velocity only matters when you're interacting with something else. The rest of the time, your speed is always zero."

"It is ok for someone as smart as your Einstein to be *almost* one-hundred percent correct," Grandin explained. "Besides, if all he said about light speed, time dilation effects, and gravity had been absolutely true, space travel would end up being really boring."

"I just wanted to make sure—wouldn't want the relativity police to pull us over and give us a ticket or anything now would we?"

Stan laughed.

"If they do, we'll just smile and say, 'Officer, velocity is an arrogant presumption—regardless of your frame of reference.'"

"Nice," grinned Claire.

***

"Mr. Garrack, thank you for coming."

President Atequipa stood and met Garrack, shaking his hand over the Presidential Seal.

Garrack pointed back at the door, "Why was I searched?"

"Standard procedure," said the President, motioning for Garrack to sit down.

"Your guards have never searched me before."

"Your nation's status with us has changed," explained the President.

"My nation's status is the same as it has always been," Garrack snorted irritably.

"Not with me," said the President as he walked around his desk and sat down.

"I fail to see the point of this meeting," Garrack said.

"Well, there have been...developments...since our last get together..."

"Yes, I know. I watch the news."

The President looked up. "It seemed necessary to compare notes."

"The only thing that I expect to hear from you is that you have apprehended the outlaws that are violating our treaty."

"If you *have* been watching the news, Garrack, then you know that these 'outlaws' to whom you are referring are now well beyond my reach." He tried to keep his voice pleasant, but it came out a mixture of pleasant and impatient; like a parent disciplining a rebellious child.

"And whose fault is that?" Garrack said sharply.

Why do I have to put up with this insect, thought the President with a sigh.

"Garrack...it was the aggressive and un-civilized behavior of your agent that prompted the CIA to warn McPherson of your investigation. Then your clumsy military had a chance to get him over the Atlantic and missed, and now I hear you missed him again in some idiotic operation in Scotland. Witness the result." He tossed a stack of newspaper clippings and magazines at the alien, half of them sliding off the desk and onto the floor. Various articles on alien fish, UFO sightings, the McPherson spacecraft, and related topics were all over the covers and pages.

"I do everything I can to help preserve the secrecy of your presence here," said the President, "and to keep my part of our bargain—a bargain that I don't even believe in—while all the time your people are out there, stumbling around, blowing trumpets to the tune of 'Space Invaders'."

"Those actions were all necessary, because your..."

"Now, Garrack, I'm of Spanish heritage. When my people first came to this continent, they fired the guns on their ships to *wow* the natives and make them think they were gods. I hope your people aren't stupid enough to think that trick will work on us."

"I don't have to listen to this," Garrack said, standing to leave.

"Then listen to this," he put away his 'nice President' voice. "McPherson is gone, and thanks to you and your little treaty I don't have the ability to retrieve him or his ship. Maybe you can find him out there, maybe you can't, but he is no longer on Earth and I wash my hands with the whole matter. It's time for you and your people to crawl back under your rock and pretend to be invisible again…oh and please recall those ridiculous fish probes. They're not fooling anyone, and I'm tired of seeing photos of them cross my desk!"

Garrack left.

After several minutes, President Atequipa opened his door and motioned to the Secret Service desk just outside his office. One of the agents, a tall young man with freckles and bright red hair, looked up.

"Come in and talk to me for a moment will you please?"

Without saying a word, the agent walked in and closed the door.

The President whispered to him. "How'd it go?"

The agent replied, "He's a git; he'll never find it."

"Even if he does, he'll think his own people planted it there—part of the magic of using their bugs instead of our own. Mills and Jones will be in touch; get to work."

"Yes, Sir."

He turned to leave; the President stopped him.

"Oh…and I'm very happy your people and mine are working together again."

"As are we."

***

"Good news, Klicktah. The Legal Committee resolved to pass the matter forward to the Security Committee."

"It doesn't matter anymore, Madam Advocate, the Zom are on the move."

"What?"

"The process has taken too long, look!" He pointed at his computer screen.

She came around his desk and stared. "What am I looking at?"

"It's an intelligence report that I just got from Gimlar. Given the time that it took this information to reach us, that fleet should already be well on its way toward Telesia."

"How do you know it's going to assault Telesia?"

"Not just Telesia. Look at the size of the force, and at who's in command."

She squinted at the display.

"Taff and Shellar."

"That's right."

"What about our little *package*?"

"It probably won't be delivered. If it's still in the system when the invasion arrives, there'll be far too many Zom ships and probes in there for anything to sneak out, stealth or otherwise. My people are going to try and seize the ship before that happens, and use it to barter some kind of deal with the Zom to keep them from escalating the conflict."

"They honestly can't think that will work?"

"They do."

"Where's *The McPherson* now?"

He looked at her for a moment.

“That's deeply classified, but I can tell you that they're at one of the large outer planets.”

“A moon is a poor hiding place.”

“Don't worry. They are very well hidden.”

“What?”

“I honestly can't tell you more.”

A sparkle of understanding lit her eyes.

“It can't be! You can build outposts on gas planets? Do the Zom know?”

“Madam Advocate…lower your voice please,” whispered Klick-tah.

“Sorry, I shouldn't have asked. You Crenum and your technology amaze me sometimes. When will the Zom fleet be detected by your people in Telesia?”

“Six or seven days before the fleet reaches them there. We project it to be very soon.”

“Then I have to contact President Shackree, this decision by the legal committee empowers me to lay claim to any evidence I want and take it from anyone. While we wait for a spot on the Security Committee’s schedule, I will arrange to apply the necessary diplomatic pressure on your people to prevent them from getting rid of that ship once they have it. Where will the Telesians go if they manage to get away?”

“The plan is Xentar or Sisna, but that's…”

“Classified, yes I know. It's good though, I have trustworthy contacts in both those systems…I can begin to feel the waters. Let me know when it's safe to start assembling investigation teams, I can’t do it without making a little noise.”

Chapter 15

"Have you found them?"

"Saunders and McPherson are still asleep; we locked them in their rooms. Standish is still unaccounted for."

"Post guards and find Standish. Check the mess area."

"We haven't checked the Telesian ship yet."

"Good…don't. I'll check the ship. If he's there, then he probably already knows what's going on."

"You can have him—Standish is dangerous."

"All three of them are dangerous," Grandin said. "These orders from the government are insane; they have no idea what they've done. Guard the other two well and try not to agitate them, and whatever you do, don't tell them why they're under guard!"

Grandin went up to *The McPherson*. Someone had already removed the scaffolding and activated the ship's systems, preparing it for flight. With drawn pistol in hand, he stalked up the ramp and slowly climbed the stairs to the bridge.

Jack Standish sat at the main computer console, writing. The energy scanner on Grandin's pistol fluctuated when he pointed it at Jack.

"I'm using one of your personnel combat shields," said Standish without turning around. "As I recall, it protects me from stun effects."

Grandin called on his com link.

"This is Grandin. I found Standish and I have him contained; board the ship."

Standish still didn't turn around, he just kept typing on the console.

"Forget your manners Grandin? Speaking Crenum in front of a Telesian who doesn't understand is rude."

"Don't treat me like I'm stupid, Jack, you speak Crenum nearly as well as I do."

"Where are Stan and Claire?"

"They're safe in their rooms. I made it clear that they are not to be harmed in any way, but they will not be released until after I've left with *The McPherson.*"

Standish looked up for a moment as the sound of running footsteps went through the ship.

"You can't fly it Grandin," he sighed, "none of your kind can."

"We'll manage."

Standish turned around. He held his pistol in one hand and a remote control in the other. His thumb held down a button on the remote.

"No…you won't. Now, I want you to get your people off this ship and bring Stan and Claire up."

"No, Jack. I have my orders."

"To Hell with orders, Glem. Those orders'll ruin everything. You came here to help Earth."

"I'm taking this ship to Crenumar. My people's needs come first."

"Listen, Colonel! They plan to sell it out! They want to try and trade it to the Zom to prevent a war; but it'll prevent nothing—it'll only give them a stronger case. The Zom don't want Earth for this bloody 'biological mining' rubbish! They want a military outpost. Do you know who the Zom are sending here Grandin? Admiral Tass. They wouldn't send a man like him to raid a place like Earth, Earth is just a stepping stone. That man is here to attack those three security outposts of yours, inside *Crenum* space. It's as plain as this big nose on my face—don't tell me you don't see it."

"Jack, I do understand, and believe me, I don't like this any more than you do."

"Grandin, the politicians who told you to do this don't give a hang about you or this project."

With that, Standish reached over with his gun hand and pushed a button on the remote. Grandin heard the ramp closing outside. The ship rose and hovered, retracting its landing struts.

A message came on Grandin's com, "Sir, we're completely locked out of the controls!"

"Get some men up here with shield negators—now!" Grandin barked.

Standish pointed his pistol at Grandin's chest and released the button on the remote. Three round panels opened in the floor, and small defense turrets—heart-stopper turrets—unfolded out of them like skinny, grotesque plants sprouting from the ground. One sprang up right between Jack and Grandin.

Grandin froze.

"Good, you recognize it," Standish said in a cheerful voice. "There are others all over the ship, inside and out. They will activate in fifteen seconds. Better tell your people.

Grandin got on his com.

"All hands onboard or standing near *The McPherson*, freeze! There are heart stopper turrets onboard! Do not move! And no one approach the ship from the outside!"

Just then, the turrets on the bridge began to emit a low hum, and flashing red lights appeared below each one.

Grandin stood absolutely still and spoke without moving his mouth.

"Jack, you are a mad man. Where did you get those?"

"Claire really is an amazing woman, isn't she," Standish replied. "Do you know, she was hooked into your alien black-market practically from the moment those cute little feet of hers set foot on this station? She probably stole something from you to trade for them. It's incredible how she can just stick her fingers into anything and pull out stuff she wants…it is truly a rare gift."

"Those are motion seeking turrets, Standish," Grandin said. "You do not know what you two have done, installing something like that on a spacecraft!"

"They can't damage the ship, they're heart-stoppers, and they only kill in a very tight, short-range beam—"

"They kill indiscriminately—lightning fast! We use them in vaults and other places where we can deactivate them before people enter. We never use them where people have to live or work. Those things can kill you and I both in an instant—that shield you're wearing wouldn't save you. You could get out into space, and something could go wrong and trigger them accidentally, and then everyone on board would arrive at their destination dead!"

"Oh, it won't shoot me, Grandin, look at it again. See that little grey box?"

Without moving his head, Grandin snapped his eyes down at it. There was an extra component on the top of the turret that he hadn't noticed before. He snapped his eyes back to Standish.

"I see it."

"Claire stole a bunch of those from a U.S. air base. It's an Infrared Imaging Module. The Yanks use them to prevent their missiles from attacking their own planes and tanks. They distinguish different targets based on their heat signatures. If the missile locks onto a friendly unit, that box shuts it off…disables it."

Leave it to Claire to hack into the brains of a Crenum weapon and get it to take commands from Telesian equipment.

"She programmed them not to shoot you?"

"No, Grandin, they're nowhere near that sophisticated, but you and I evolved under different stars—our thermal fingerprints…I think that's what Claire calls them…are different. She programmed these turrets to distinguish Telesians from anyone else—here, I'll show you."

He waved at the turret. Like the strike of a snake, it spun around and pointed at him. Then it seemed to power down and reset. Two seconds later the red light was blinking and it was scanning the room again.

"See? A bit clumsy, of course, but reliable enough. *You'd* best not try it though, my friend, it'd kill *you*."

Grandin slowly and carefully took another breath. "I am pointing a stun weapon at you, *my friend*, and you are pointing a firearm and heart-stoppers at me and my people."

"I guess that's the difference 'tween you and me, Glem ol' boy."

"We were helping you."

"That's correct, you *were* helping us—right up until this morning. I guess we've just out-lived our usefulness to one and other. Listen, this is a lovely chat, but it's taking too long. There are, what, ten or so Crenum aboard at the moment, not counting you? Maybe a few standing outside? I like you and your people a lot—I really do. I'd truly regret it if any of you sneezed right now. So here's the deal, get on your wireless there and have

Stan and Claire sent up. I'll land the ship, drop the ramp, and once they're both aboard I'll deactivate the turrets and you and your people can get safely off."

Grandin thought for a moment.

"Look, Grandin, we're wasting time. You have to evacuate this system before the Zom get here and blockade it. They're probably, what, to the edge of the Oorte Cloud by now?"

"There's nothing I can do, Jack..." Grandin hissed between his teeth. "...the turret."

"Oh, sorry. Forgot." Standish pushed a button on the remote and the turrets on the bridge shut off.

"Careful what you say into that wireless, Grandin, I do speak Crenum...and Zom. I've missed a lot of sleep lately, helping Saunders install those turret things. They're complicated and evil tempered, and they've left me in no mood for tricks."

Grandin activated his com, "Release McPherson and Saunders. Bring them quickly to the dome level. Send them aboard alone, *do not* approach the ship."

Jack smiled.

"Good boy," he said. "And once all that business is complete, we'll be leaving straight away. Don't try to stop us either...all those nice weapons you installed for us could make a ghastly mess of things."

***

"What in blazes...has everyone gone mad?"

"A Zom fleet is coming to invade the system, Stan," replied Standish. "These people are scared."

The dome started to fill with liquid methane.

"Raise shields," said Stan. "Claire, get below and put things in order, I seem to recall the Crenum haven't finished their work down there yet."

"I'm on it."

He looked over at Standish.

"Well…thanks. I didn't see it coming."

"No problem, that's what spooks and thieves are for. So are we headed for Sisna?"

"Let's just worry about getting off the planet first. Pull up the safe escape data that we collected on planet-fall."

Crenum ships arrived outside and removed the dome segments as soon as it had filled, and the McPherson lifted off from the tower and headed for the surface.

"Any moons or rings above us?"

Standish checked the nav computer.

"Nope, but might I say I do like these new computer upgrades."

As they broke the surface, the ship stopped and hovered.

"That's odd. Claire, how are things?"

"Well, they could be better. A lot of the connections between the ship and that fancy new alien nav computer up there on the bridge are still all mixed-up in a bird's nest of wires on the floor down here."

"Great, so what do we do?"

"They dumped the old bridge nav computer down in the hold; it's too big to haul through the ship to the bridge, but I think I can drag it in here and plug it into the service port on the drive control unit."

"That'll get us off the planet safely, but we need the alien nav computer hooked up in order to leave the system, ours doesn't know where Sisna is."

"I understand, but it'll take a while to figure out the connections to the alien unit.

"Standish, we have to warn Earth about what's coming...do your people at Section Five have an alien transceiver?"

"Already ahead of you, Doc, I wrote them a message while I was bantering with Grandin earlier. It's short and to the point, only four or five nanoseconds long and in code. Just say the word and I'll send it."

"I don't want to risk the Zom tracing it back to our friends here. Wait until we get away from the planet a bit. We can give them details when we get to Earth."

Standish looked up.

"We're going back to Earth?"

"We're leaving the system, Jack, but first we need time to finish getting these upgrades connected—and tested. There are also a few people and things I wanted to take with us when we go. I originally designed this ship so I could fly it alone, but it's way too complicated for that now. We need a crew—and we just left most of our hands back there on Saturn Base."

"Jack," Clair said, "come down here and help me with this beast."

After Jack and Claire got the old navigation computer going, the ship rose swiftly up through the wind and clouds.

Jack sat back down in his seat on the bridge, and then whistled as he looked at his display.

"Stan my friend, bad luck. Two Zom ships are moving into our path, looks like a super transport, and a fighter pod."

Stan checked his computer.

"I didn't know they had anything that big in the system."

"It can only be their administrative transport. The Zom Governor's ship. It's a refitted troop transport. He uses it as a base of operations."

"Kind of a city hall in the sky."

"Ya, basically."

"Looks like that pod has all five of it's fighters too."

"Well, Stan ol' buddy, you wanted to fly in space someday."

Stan thought for a moment.

"We can't maneuver out from under them until we reach our gravity threshold, and they won't know what that is."

"Umm," Claire cut in over the com, "they can guess pretty close for their purposes though."

"Claire, can we take them?"

"Take them? A bit ambitious aren't we? That transport would have a plasma missile launcher on its stern...and lots of particle beams, and it's also got a few dozen troops with armed shuttles and particle drills for boarding us through our shields. The fighter pod can use the missile launchers and particle cannon from its fighters, but it has to be facing us to do it. Without the fighters, it's unarmed. Both ships are shielded roughly the same as us. With this new armament, we could probably beat the pod and its fighters, or the transport and its troops, if we can keep our distance. Both ships at once though—I donno. They can swarm us with small craft and force their way aboard. Either way, a serious fight here will take time and they can call for help—we can't."

"What I want to do is just leave and outrun them."

"Nope, they'll kick our butt. Both those ships, and their fighters and shuttles, have two to three times our acceleration. We have to get far enough away from them to disappear from their visual sensors, or they'll just chase us down and force us to fight."

Just then, Standish got a message on his console.

"You know, Stan, I am getting so sick of Zom arrogance I could puke. They're telling us that this ship is Zom property and they're ordering us to approach the transport and stand-by to be boarded once we clear our gravity threshold. They're turning that transport's belly toward us to block our path, and the pod is turning to face outward—to be ready to chase us down if we try to run."

"Can they see our new weapons yet, Claire?"

"Apparently not, with that stupid arrangement," said Claire. "This changes everything. Nobody's pointing any missile launchers at us at all now, and if we wanted to we might be able to just toast'em both and they wouldn't stand a chance."

"They'll regret that particular positioning of their ships, that's for sure," said Stan. "The transport's so large it's completely eclipsing the pod, so we can go after them one at a time. I don't want to squander missiles though—we can't replace them. Besides, if we kill everybody, with Crenum weapons, the Zom might think the Crenum did it. That'd be bad for our friends later—we need witnesses."

"Ya, and we can't just keep assuming that the Zom are nit-wits either; they aren't," said Standish.

"Jack, look at them!" Claire shouted. "That transport might as well have 'Kick me' painted on its belly! They have no clue what they're in for—they must think we're still armed with Earth weapons."

"We're not going to kill the transport if we can help it, Claire."

"Why'd the Crenum give us those weapons if they didn't expect us to use them?"

"Claire, they gave us the weapons to defend ourselves, not to slaughter civilian government employees on transports. Now, do you suppose those people have worked here in Telesia for a while, like the Crenum on Saturn? They'd have their families with them, right?"

"I suppose so."

"How many missile hits can that transport take?"

"No more than six, tops."

Stan looked up at it and wondered if Clell was aboard.

***

"And you think that after all that's happened, they'll just throw up their hands and surrender?" Clell couldn't believe what he was hearing.

"What else can they do?" said Governor Thrigh. "They don't have a choice; they can surrender or run, and we have them blocked. They can't fight us, and they can't maneuver until they reach their gravity threshold. If they do somehow get around us, the pod can just overtake and disable them."

Something about the situation bothered Clell.

"I still say we should just let the pod unload missiles on them and knock them out of safe ascent; let the planet kill them."

"We need to recover more from that ship than charred, wadded up wreckage," Thrigh said with a huff. "There'll be evidence of Crenum tampering onboard that ship...maybe a few Crenum too...that'll help justify a war with Crenumar."

We shouldn't be starting a war with Crenumar, thought Clell. *The Telesians are bad enough.* He felt the scar on his cheek again. He'd told the doctors to leave something of a bulge underneath the skin—for a reminder of how stupid he'd been. McPherson had shot him with a gun that he'd produced practically from nowhere. Saunders had hurt him when she'd escaped too, and he'd thought she was helpless. Yet both of them were sheep compared to that hyena Jack Standish. All three of them were likely aboard that ship and everything in Clell's being told him that killing them, quick and clean, was the only safe way to handle them.

"I think you're underestimating these people! Shoot them down!" Clell yelled.

The Governor turned in his chair.

"What?"

"Shoot them down!"

"Don't give me orders!"

"You're making a huge mistake! I know these people."

"Know them? If you'd killed McPherson in the first place, we wouldn't be in this mess!"

"Yes," Clell admitted, "I underestimated him—just like you're...

"They've just sent an acknowledgement to our order," said the communications officer.

"See, that? They're surrendering after all," said Thrigh. "Just in case, have a boarding shuttle with twelve Marines prepped for flight and standing by."

"They're painting us with graviton beams."

"That's fine," chuckled Thrigh, "their weapons can't hurt us."

The weapons officer spun around, his eyes wide, "Sir, they've just fired four plasma missiles."

"What? Activate counter measures! Whose technology are those missiles? Not Telesian."

Clell groaned.

"No, it's Crenum anti-piracy armament, Sir, the data's coming in now...they class somewhere between a patrol ship and a light destroyer. Sir, the data we had on them...!"

"They seem to have had an upgrade," Clell grumbled.

Into the Dark

Escape of the Nomad

"All ahead full!" shouted Thrigh. "Give the pod a clear line of fire! MOVE! Where are those countermeasures?"

Their primitive, helpless prey had somehow grown teeth sharp enough to destroy the ship…

…and it was on the attack.

A large piece of the Zom government in Telesia was based here. In addition to the crew compliment of fourteen, the cargo bays and most of the troop barracks provided office and living space for over two hundred persons and their families. In addition, most of the information and infrastructure that they, and the approaching invasion fleet, would need to coordinate the conquest and early management Telis 3 were here. They were too close to Telis 6 and they were not in orbit.

He glanced down at the churning atmosphere of the planet below. If the ship were to lose propulsion and fall below its gravity threshold…

***

The McPherson increased speed, following the missiles in. Particle beam "pulse" energy from the transport splattered against the shields, throwing sparks as if from a blacksmith's hammer. Each hit chiseled a tiny, visible mark in the shields that would last for hours. Shield generators could only repair particle weapon damage slowly, so even though the damage was slight, it accumulated. Without shields, any particle beam hit would pass completely through the ship, leaving a pin-sized hole.

"They're launching counter measures and trying to move out of our way," said Standish.

"Too little too late, the missiles are already there. Jack, send that message to Earth."

They watched as all four missiles hit the belly of the transport simultaneously, bursting open and engulfing the craft with blue and yellow fingers of energy for several seconds. By the time *The McPherson* reached the two ships, the transport had moved and they could see the fighter pod finish its turn, weapons at the ready. More sparks exploded over *The McPherson* as the pod, too close to the transport to use missiles, opened fire with particle beams.

"Jack, return fire, but only against the transport. Try to focus on the drive ring."

"Ok, but don't expect much for accuracy, I've never used these things before."

Stan altered course slightly to bring *The McPherson* closer to the transport, while Standish lit up its stern with particle energy. The pod, unable to un-dock its fighters while maneuvering, continued to keep its guns trained on them.

Then *The McPherson*, after accelerating past the two ships, fired four more plasma missiles ahead of the transport.

***

That was enough for the captain of the fighter pod. He could hear his pilots sitting in their fighters, openly cursing him for not stopping his turn, but he ignored them. Other heads would have to roll for the *Cladnic* escaping again, his primary responsibility now was the safety of that transport and the people aboard. It was up to him to prevent a catastrophe. He matched headings with the transport and punched his acceleration to maximum, to try and interpose his undamaged shields in front of the missiles. As he did, he launched countermeasures, angling them so that they would lure the missiles away from the transport and toward him.

One of the missiles struck the bow of the transport, shattering what was left of her shields. Two of the missiles trailed off, chasing the fake readings from the countermeasures and detonating far below in the thick atmosphere of the planet. The remaining missile turned and struck the fighter pod, sparing the damaged transport from further hits.

***

"That pod just launched four of its fighters, but I can't see them," said Jack.

"Good, let's hope they can't see us either," said Stan. "If you see a fighter, blow it away; don't wait for orders..."

"Stan!" shouted Jack, spinning around in his chair and drawing his pistol. "The docking port!"

Stan turned, just in time to see the door fly open. Twelve large Zom in some kind of light battle armor charged through the doorway and onto the bridge. They pointed their weapons at Jack and Stan.

"Klifelley!" their leader shouted.

Jack nodded.

"Claire, did you hear that?" he said, as he carefully set his pistol on the control console. "There are Zom Marines up here. They want us to *Klifelley*—to surrender."

"Ok," said Claire, with a sarcastic lilt in her voice...

...and the three heart-stopper turrets on the bridge deployed.

"Rect! Sheffel toom!" shouted the leader—and then he fell dead.

The bridge burst into cacophony of falling bodies, flashing weapons and exploding turrets. Jack grabbed his pistol and fired through the smoke, striking the last remaining Zom in the transparent faceplate of his helmet.

"Trass, " the Zom chuckled. "Tell sraft id klifellod." Stan could see the flattened bullet wedged in the Marine's faceplate, directly between his eyes.

"Well," said Jack with a shrug, "can't blame a bloke for trying."

Then the Marine gasped. His eyes went wide and his knees buckled under him as he fell forward onto his face. A wisp of smoke hissed from a smoldering red hole at the base of his skull.

Claire Saunders stood behind him, her pistol drawn.

"Jack," she said with obvious annoyance, "when are you *ever* gonna get a decent weapon?"

***

CIA agent Rod Mills charged out of the doorway of his apartment at a run, a suitcase and a briefcase clutched in his hands, shouting into his cell phone earpiece. He ran down the stairs, hitting the banister painfully with his hip on the way down, and jumped into a waiting taxi.

"Yes! Now! I said NOW!...I don't care what time it is. Get him!...Well, if he's already on his way home then bring him back!...I don't care about his son's baseball game...give me his cell number...Your job!? Do it on my authority, trust me, we may all be out of a job in few days anyway."

He called several other people, and several others called him. As he talked, the taxicab driver kept looking back at him through the rear view mirror, concern etched on his face. The taxi stopped at CIA headquarters and Mills jumped out with his things, but when he turned back to pay the driver, the taxi had sped off. He watched it leave.

Must be heading home to get his family out of the city. Good for him.

***

Withim walked into the President's office and handed him a note.

"Sir. They're coming."

"How many?"

"We're not sure, but the larger ships aren't hard to make out, even in Neptune's orbit. There seem to be four core groups that took up orbits around Jupiter, Saturn, Uranus and Neptune—initially—then the Jupiter group left orbit and is now on its way here."

"Attack as soon as that group is in range."

"Yessir."

***

On a remote road in the Arizona desert, the five-ton truck pulled up to an extinct filling station, at an intersection between long roads going nowhere, and stopped. Two old-style gasoline pumps adorned the property, along with a small office/store and a mobile home. The driver looked at his watch, 3:35 am, a little early. He climbed down quickly, eager to unload and return home to bed. He was lowering the ramp on the truck, when the door on the mobile home opened. A kindly old man with a broad smile leaned out.

"Just leave the delivery there by the side of the road! I'll be out in a minute."

The driver climbed into the back of the truck and started unloading the three pallets, using a grav-lift that was inside with the cargo. By the time he'd finished, the old man had dressed and walked out to sign the delivery voucher.

"Leave the lift vehicle please."

The driver checked the invoice; sure enough, the grav-lift had been included with the purchase. Someone must have a lot of money. He shrugged and handed the old gentleman the keys and paperwork.

"Are you married young man? Got any family?"

"Nope. I'm alone."

"Come in and have a cup of coffee with me—something very important is about to happen, and I want to talk to you about it. I promise you will not regret the time spent."

"Ok, sure."

He followed the old man into the house.

***

Dr. Howard Turnblum sat in his basement, studying the stars from an old database. He was comparing the time-spans between supernova remnants and new star formations near them, looking for correlations. It was very old science, but repeating and verifying old discoveries was better than complete boredom.

His cell phone rang.

"Hello, Howie," said Stan McPherson.

***

"Well, it all seems to have tested out ok," said Claire as she fell into her seat on the bridge. "I still think we should try a shorter, slower jump first though…say, out to Neptune, before we try and jump to Sisna."

The McPherson flew east, low over the Atlantic, toward their hiding place near some geothermal vents in the transatlantic ridge. Earlier in the night, they had delivered the Zom attack shuttle, along with it's dead soldiers and most of their armor and weapons, to the U.S. Naval base on Diego Garcia. Later, they'd dropped off Jack in Virginia, to meet with the CIA about the coming war with the Zom.

"We'll see what the Zom let us do. They might own the whole system by the time we get out of here at the rate we're going," said Stan. "It would have been nice if Grandin's people could have had one more look at the systems."

"Grandin," Claire said. "So much for friends."

"Now, don't take him off your Christmas list, Claire," said Stan. "Like Jack said, he had orders direct from his Governing Council. He did what he had to do."

"But to sell us out to the Zom!"

"Not us, Claire…just the ship. Nobody ever proposed that they hand over people."

"How does Jack know so much anyway?" Claire said. "Those orders would have been secret, yet he seemed to know what they said, almost at the same time Grandin did."

"Ya," said Stan, "and he speaks Zom too. Look, he's not the only one here with shady alien connections—where did you score alien heart-stopper turrets? I doubt they just give those things away with a kiddy meal at the local stop and choke."

Claire laughed, "Maybe they do! I'm not telling though—a thief has to keep her secrets."

***

The U.S. Navy SEAL team arrived at Arecibo Radio Observatory just after dusk, securing the facility and its immense radio dish easily—

through prior, peaceful arrangements with the appropriate authorities stateside. Select members of the operations staff were called in to help, and in a few minutes they had hooked up and tested the portable computer console that the SEAL team brought with them. Only four men there really knew all of the details of what they were about to do. It had been done only twice in human history, and generations had hoped that such action would never have be taken again.

***

"Admiral, one of our patrols captured a Crenum shuttle enroute to Telis 3."

Once Admiral Shellar had left the vaunted Admiral Tass behind to secure the outer system, he headed straight for the inner planets. He wanted to get his transports landed quickly, in case the Crenum changed their minds about interfering.

"Confiscate it and have it searched."

"Already underway, Sir. Captain Thrigh requests that the shuttle and its crew be transferred to his ship for search and interrogation. It seems the ranking officer aboard the shuttle is a Col. Glem Grandin, who has been very active in the Crenum espionage of this system. We've already found some equipment and data aboard the shuttle that was used in the construction of the *Cladnic*."

"Give nothing to Governor Thrigh," said the admiral.

"Yes, Sir."

Right after I control the planet, I'm sending the bungling fool home anyway.

"Hold the Crenum personnel under guard aboard this ship, as our *guests*. Just because we have to take their shuttle away from them, doesn't mean we need to be uncivilized about it. We aren't at war with the Crenum—yet. When we are, then we can move them down to the brig and question them properly."

***

"Mr. President."

"Come in, Harold, tell me some good news."

"Only a fraction of the two thousand X-13 units were ready in time. We're servicing the others, but they won't be part of this launch. The special ops teams have all arrived at their destinations."

"Activate the system, launch the weapons."

"Yes, Sir."

"Funny how things work, isn't it, Harold? We take one photo of an alien fighter, see a passing resemblance between it and an old, rejected NASA design, and before you know it, you have a weapon."

"Well, it's pretty iffy as far as weapons go, Sir. Personally, I think any commander who gets his units killed by something like this would have to have the arrogance of a grizzly and brains of an amoeba."

"We have to try, Harold, the Zom have worn out their welcome."

"It's the lag time that bothers me the most...but if the enemy allows the system to fully deploy, it'll be a very loud bang—one of the loudest in Earth's history."

Chapter 16

"Give me a scanner sweep of the target, Science."

Admiral Shellar was proud of his reputation for crushing defiant military systems, rapidly and ruthlessly taking over while leaving the planets' political, economic and industrial infrastructure intact—for the most part—and handing over a functioning planet to his superiors. He wished there was a faster way to approach the target though and felt relieved that he could finally get on with it.

"Sir?" said his science officer, Lieutenant Commander Mel Janchet. "We're detecting an aberration. There are over five hundred more artificial satellites orbiting the planet than there should be."

"Noted," he replied to her with a wave of his hand.

These were probably just old leftovers, derelicts that the intelligence types hadn't considered noteworthy. With so many satellites, a discrepancy of a few hundred was typical, and easily explained a dozen different ways. Of course, it was his crew's duty to be alert to such things. Once they had saved him from a nasty embarrassment when a foreign freighter had accidentally passed too close to one of his formations. The law would have justified him in destroying it, since it could have just as easily been a suicide bomber. Janchet had thought to scan it first, and found children aboard, so he'd handled the situation differently, boarding the ship and arresting the officers instead. That's why he allowed, even encouraged, the crew of his flag bridge to note anything out of the ordinary—then he'd decide whether it was noteworthy or not.

"Sir," Janchet spoke up again, "it seems that those extra satellites are some kind of probe. They're leaving orbit and heading this way.

"Pull up the intelligence report on this planet; explain those craft."

"Sir, the planet has some ability to launch extra-orbital probes, and possesses primitive versions of a few advanced weapons."

The admiral grunted and turned to his tactical officer.

"Do those probes have shields?"

"No, Sir."

"Weapons?"

"No, it looks like they match a new type of inter-planetary probe that's listed in the database as 'experimental'. *Racthra* is requesting permission to launch a fighter squadron to disperse them."

"You mean attack them? No. Let their probes have a look at us. Continue as planned, but order the fleet to activate shields, just in case.

"Sir, they're coming in awfully fast for grounded spacecraft," said Tactical.

The admiral thought about that.

"Science?"

"They seem to be using artificial-gravity propulsion."

"Let's see one."

An image appeared on the viewer, showing the front of one of the diamond-shaped craft. A raised area encircled its middle and solar panels made up its skin.

"Is that bulge a drive ring?" the admiral asked, squinting at the view screen.

Janchet zoomed in on the bulge.

"It could be. Looks like it's partially concealed under the outer skin.

"Yet they're definitely not Crenum."

"No, Sir."

They obviously carried no danger to his armada, but artificial gravity technology would give the UCP authority to interfere with Zom plans for Telis 3. If he looked at these probes too closely, and verified that the propulsion system did indeed include a drive ring, he might have to answer difficult questions later. He wanted to be able to say that he had assessed their tactical threat to the landing and found them toothless. If they constituted a political issue, that was someone else's problem.

"Order the fleet to ignore them."

The tiny probes swarmed them like smoke until they had surrounded half his ships. The armada continued to slow as they drew closer to Earth, with the new tagalongs matching their course and speed.

***

The SEAL team commander's cell phone rang. He already had it in his hand and answered immediately. "Yes? Yessir, Mr. President." He hung up. Two of his men also received phone calls.

"Gentlemen, I have a coded order to execute," the commander said, his voice cracking a little. "Do you concur?"

Both men nodded.

Without hesitation, each of the two men stepped forward and produced a flash-drive from a cord around his neck. Their commander plugged each into the portable console, then turned a key and typed a pass code. A large red button on the console lit, and flashed. The commander watched it blink as the seconds ticked by. He started to say something, and then changed his mind. With a sigh, he shook his head and gazed off into a corner of the room for a moment, then casually pressed the button down with his thumb. The light went solid for several seconds, and then went out. The operator monitoring the transmission turned to the commander and nodded, indicating that they had sent the signal.

All in the facility quickly, quietly, almost with reverence, packed up and left. The observatory staff went to tell the local authorities what had just happened—-and the SEAL team vanished into the jungle.

***

"Sir?"

"Yes, Lt. Commander?"

Janchet paused a moment.

"Never mind," she said.

"Go ahead, Mel, what is it?"

"I really think we should deep scan these objects, something doesn't feel right about them."

"There are several reasons why I don't want to look at them too closely, Lieutenant Commander."

She turned in her chair to look at him.

"I understand, Sir, but I don't like their spacing pattern—it has an eerie…mathematically precise…engineering to it. I think they may pose a danger to the fleet."

Those were the magic words. Regulations required him to listen if a subordinate had safety or security concerns. If he ignored her, and something happened, there would be repercussions later.

"Go ahead."

She looked the scan over, but the puzzled expression on her face remained.

"Well?" he asked.

"I can confirm the drive rings sir; they seem to be powered by hydrogen fuel cells and solar panels—very primitive, definitely not Crenum. Strange…the shape of the drive ring almost looks like something *we* would use, but the technology and the construction materials resemble plans that the Telesians worked on fourteen stellar cycles ago, and then scrapped."

Just what I was afraid of. We're supposed to be pretending they never developed that.

"I'm already familiar with it, Science; it was the precursor design to the *Cladnic*."

"Yes, Sir. Something odd though; these devices also have heavy metal radiation shielding inside them, surrounding storage tanks full of fusion fuel, but I can find no fusion reactor."

"Then what's wrong?" the Captain asked.

"One moment, Sir. There's something about the internal portion of these probes, near their fuel tanks." He could see her flipping through Telesian weapon designs on her computer console.

"Talk while you work, Science. Let us all in on what you're doing, please."

She put her data on the main display for all to see. It showed the positions of the probes around the ship and close-up cross sections of her scans, overlaid with various known Telesian technologies.

"These craft are quite massive for their size, Sir, with a small core of heavy radioactive metal. Radioactivity is often used by primitive cultures as a power source for long-range space probes, but not like this. The core is surrounded by chemical explosives, which could be part of a self-destruct system…I don't know. Plus, the fuel tanks each contain a separate, radioactive, heavy-metal body."

The bridge was quiet as they all considered this. Finally, the science officer found the data she'd been looking for, quickly cross-referencing it with some more specific data on older Telesian weapons.

Then she turned to face him—the blood had left her face and her eyes were wide.

"Sir, the probes' outer design is a ruse," she said quickly. "Those things are multi-stage fusion warheads!"

The admiral's bushy eyebrows furrowed as he looked at the display. It was all there. For many hundreds of years, his people had used artificial gravity compression to ignite nuclear fusion reactions. Only at the very earliest stages of military training was fission reactive ignition even explained. The heavy metal bodies inside the probes were fissile material, ignited by the chemical explosives. These in turn ignited a second nuclear fission reaction, which in turn irradiated and compressed the fusion fuel, resulting in a third stage fusion explosion.

He and his crew knew that the primitive worlds they were usually assigned to "deal with" used that technology, but the savages usually just lobbed such weapons at them on chemically propelled rockets. Their shields could easily deflect whatever few might somehow manage to escape the pickets' weapons and pose any real threat to the fleet.

Not this time, he had let them all come in nice and cozy. He began to see what the science officer had noticed earlier on, about the positioning of these "probes". The warheads' formations seemed designed to implode his ships' shields from all directions, with nowhere to deflect the energy.

"Order the carriers to launch fighters..."

"Sir, there's a transmission coming in from a large EM dish on the surface of Telis 3. It's a numeric code, followed by what looks like a time stamp."

The admiral jumped to his feet and pointed to the tactical officer, who was already sending commands to the fleet. The admiral gave a loud command—

—but no one was alive to hear it.

***

Back on Earth, the men of the SEAL team looked up through the forest canopy as a bright flash, like a second sun, briefly appeared in the morning sky.

***

When the sphere of light, superheated gas, and very small pieces of molten debris had dissipated, half of the invasion fleet was gone.

Admiral Taff looked on from the bridge of his flagship, in orbit around Telis 6, and watched as the surviving vessels of the late Admiral Shellar's fleet continued toward Telis 3. He shook his head as he viewed the damage and casualty reports that flashed across his console. His government had sent people to this planet to learn—and short circuit—their technology, to prevent this sort of disaster. Their people had been sending intelligence reports for fourteen Telesian cycle. How had they failed to notice the development of such weapons?

There were not supposed to be *any* indigenous space warfare vehicles of any kind here, much less defensive suicide drones equipped with drive rings.

If the UCP ever found out that this world, a world that was supposed to have stagnated itself, had developed weapons equipped with artificial gravity drives…

“Those were Crenum weapons,” he announced.

His flag crew looked up at him briefly, and then looked away. No one appeared interested in arguing with him about it.

“Who's the ranking commander over there now?”

“Captain Gibrack, of the carrier *Racthra*.”

“Tell him to comm me.”

***

“What did you say?”

“I said the invasion is here…but something's wrong.”

“What?”

“Something major…big. I can't put it together. To much…

The outer wall of the office suddenly disappeared in a cloud of noise and dust. Armed soldiers in masks and hoods charged in through the hole and started shooting people with pneumatic darts and firearms.

“Look out! Get your personnel sh…”

***

“Mr. President…it's begun.”

“How are we doing?”

“Half the approaching space fleet appears to have been destroyed in the nuclear strike, and we've nearly completed the round up of Zom operatives. We seem to have achieved complete surprise in all operations so far.”

“Good, have we captured any useful technology?”

“Yessir, two spacecraft, a number of personal shielding devices, alien small arms of various types, and a lot of communications equipment and spare parts. The Brits are helping us sort it all out but they're stumped on some of it."

"What's the rest of the Zom fleet doing?"

"They stopped. They seem to be exploring their options."

"Agent Standish, can you help us identify some of the captured equipment?"

The Brit smiled.

"I know someone who could, she's a yank who's been working with the Crenum for a while."

"Would that be Ms. Saunders?"

"Ya, but she might need some…assurances."

"What kind of assurances?"

"The 'immunity' kind."

"We'll see what we can do."

***

"Stopped the invasion?"

"They went back to Telis 5—what was left of them. They appear to be awaiting reinforcements."

Gimlar smiled and shook his head. A so-called defenseless people had dusted half the Zom invasion force, in practically the blink of an eye. Hadn't he said that Telesians were dangerous enemies?

"We've confirmed that the Telesian suicide drones were equipped with McPherson type drive rings, but they differ somewhat from the one on the McPherson spacecraft," the Crenum Supreme Commander said. "The technology they used is clearly a derivative of their original NASA designs, but the shape of the drive ring suggests that they tried to copy the drive ring on Zom shuttles. Nobody knew…"

"What about the blockade force?"

"They left."

"Left?"

"Jumped out right after it happened. We think that…"

Gimlar turned to his console and began to type quickly. "I need two favors from you," he interrupted, "one of which I'm sure you will be very much opposed to granting me." He worked as he spoke, a grim smile forming on his face.

The Supreme Commander raised his eyebrows.

"Gimlar, I think that Tass is going to attack our outposts. Our strategy teams put together two responses that the Zom might take after a disaster of this magnitude. One projects that Tass would assault the planet himself, more carefully than Shellar. The other scenario has Tass fabricating an excuse to assault us and try and put Telesia deeper inside Zom space."

"Then we now have a full fledged war on our hands," said Gimlar. "I agree with you, I think that blockade force is enroute to Crenum space. I want you to break loose some ships to go hit that invasion force Tass left behind in Telesia, the moment you see his fleet enter our space."

"If I do that, then the Zom fighters in Telesia will need to keep close to their fleet, for defense…"

"…and reduce the damage they do to Telis 3. Exactly! I also need you to declassify that intelligence report that you have there in your hand, the part that describes those Telesian suicide probes. I want to release it to the press."

The commander stiffened. The report was gathered with the help of secret sources, ways and means, which collected intelligence in Telesia. Releasing those reports meant revealing some of the features of the stealth probes that the Crenum military used there. Letting the public see that report meant letting the Zom see it, and that would compromise the positions and capabilities of those probes, and hint at some of the communications and concealment tricks they used.

"Well," he sighed, "the two probes that transmitted this were destroyed by the Zom; I can give you their data, but the communication stations that they transmitted to will have to be moved…"

"I understand. What about *The McPherson*, did they get away?"

"Well, yes and no." He looked up and his eyes told the story.

"McPherson returned to Telis 3?"

"Yes, he did. Grandin tried to prevent it, but he was out maneuvered."

"Was anyone hurt?"

"No, but there could have been. That Standish fellow and Saunders secretly installed heart-stopper defense turrets on all decks of the ship and then used them to hold Grandin and his entire boarding party hostage when they tried to seize it. Told them to either die or let the ship go."

"I wish they'd just sent McPherson straight to Sisna," said Gimlar, shaking his head. "Now see what we've done? Now, with Zom fighters running about the area, blowing things up, McPherson's going to be trapped on Telis 3. Do you think his ship has been captured or destroyed?"

"No…but I'm not sure how we'd know if it had. The Zom would never announce it, since they're in the business of pretending the ship doesn't even exist."

"Do you have the ability to contact him?"

"We can call him directly, or maybe send him a message through our contacts on the surface, but it's not a good idea. We're not sure the Zom even know they're there. We might give them away."

"I'll be discrete."

"Also, after what we pulled, they might not trust us anymore."

"Ya, well, can you blame them? We may not even need McPherson anyway, now that we have these new developments. What do you think would happen if the Crenum people and our allies found out that the Zom attacked that planet *after* the Telesians showed them a weapon system equipped with a drive-ring? Weapons advanced enough to damage and destroy modern fleet units—weapons that the Telesians developed on their own, with no help from us, but which show evidence of exposure to Zom technology? A Zom influenced adaptation of a space propulsion technology that the Zom claimed the Telesians had completely abandoned?"

"The public would be outraged. It would effectively place the Zom in direct and obvious violation of both the letter and spirit of the Mistaac Accord, even more than the existence of the McPherson craft does. I see where you're going with this."

"Good."

"I'll have a cleansed version of this report sent up to you within the hour."

"Excellent," said Gimlar. "I'll arrange a press conference, and see that Mr. Klicktah publishes a press release simultaneous with ours. We will provide a copy of your report to both the UCP and the press for the masses to judge. Tass will attack our outposts—there is nothing we can do to stop him from doing that at this point—and that will frighten our legislators into growing some backbone...funny how that works isn't it. We'll attack their invasion fleet in Telesia as a counter attack for their assault on us." Then he grinned again at an afterthought. "And if McPherson is still in the system, maybe he can slip out amidst the confusion of an all-out border war. What do you think about that?"

"Icing on the cake."

"Exactly," said Gimlar, thumping his desk.

"I'll order Admiral Zelliut at the border outposts to respond to any Zom incursion with an immediate attack on Telesia. It won't take much—the Zom down there look pretty battered as it is. Our probes say that that invasion fleet lost two of its three carriers in the Telesian attack, and they've got a lot fewer cruisers and destroyers now than before. We'll hit them with a few assault cruisers, that ought to keep them busy enough. I can't spare any more than that from the border outposts right now, not with a man like Tass going after them. It'd be too risky. We'll soon be at war with a star nation twice as powerful as us—and if we lose those outposts then Admiral Tass, the Zom's most dangerous officer, will be within long-range striking distance of that chair your sitting in, Gimlar."

"Did our people on Telis 6 get out?"

"Most of them, they jumped for Crenumar before the Zom took the system jump points, but Grandin and some of his people stayed behind."

"Well, let's hope he keeps himself out of the Zom's way."

***

Once again, Stan McPherson was alone.

His ship flew on autopilot through the night sky, low over the dark waters of Hudson Bay, toward the polar ice cap. His hands were busy rebuilding one of his tiny probes, upgrading it to include an alien power source, stealth-shield generator, and sensor package.

Recent events had left him frustrated—so many things for a mind to take in at once. He hadn't returned to Earth intending to help in its defense. He'd planned to be far away before the Zom arrived. Yet here he was, hiding from the Zom again, only this time from the hundreds of fighters sent to ravage the military assets of the United States and its allies.

He'd known that Jack would need to coordinate with his counterparts in the U.S., but he'd never dreamed that Claire Saunders, who'd so openly considered herself more alien than Earthling, would want to risk her neck to go to Washington and share her knowledge with the people who had once helped the Zom hunt her.

Now, instead of evacuating the system ahead of the Zom, like any smart person would do, he and his new starship had to hide under the ice and wait. He couldn't even collect the people and things that he wanted to take with him—his food supplies were limited and he now had no idea what the future held or when, or if, he would get back into space again.

He could leave now, like he'd planned to do in the first place—back when his life had been simpler. He could leave behind the problems of Earth, and all this interstellar intrigue, and lose himself in the heavens. The U.S. Government—who until very recently had been in bed up to their necks with the Zom—had chased him away to the British Isles. He owed them nothing. The Crenum—his new friends—had tried to sell his ship to the Zom in exchange for peace.

Right now, he wanted nothing more than to just find some isolated corner of the vastness of space and vanish into it. The Zom ships were positioning to defend the gas giant planets in the system—and what was left of their invasion fleet—from a possible Crenum strike, but that didn't necessarily mean that the Sol system was closed. Proxima Centauri was useless to the aliens, uninhabited and ignored, and it was near enough—he didn't need to jump from a gas planet to get there. It might take a year or more, but he could make it. There was enough food and fuel aboard right now to easily put the Zom, and Earth, and everything else behind him and just slip away, unseen, into the dark.

Interstellar space lay opened to him.

But it really wasn't that simple anymore.

Into the Dark

Escape of the Nomad

He had changed.

He had become more than just lonely former astronaut Stan McPherson. He had a family now, people who relied on him, people who needed him and trusted him, people who had lifted him from the depths of loneliness, despair, and desperation to a man of respect and influence among them. The Crenum had helped him finish his ship, and now they needed him to fly it to the fifth planet of an alien star they called "Redliv", the core of a star nation that was neutral to the tensions between the Crenum and the Zom. They needed him to show his ship to the people of that world, and to the interstellar community. The Crenum hoped that by doing so Stan might save them, and Earth, from a protracted war with the Zom.

The President of the United States had spoken to him personally and apologized for Stan's past treatment at the hands of the government. He had asked Stan to stay on Earth a while longer, while the White House compiled a package of evidence of Zom crimes against Earth for Stan to take with him to Redliv 5.

Claire Saunders needed him too. She'd given him instant acceptance, and friendship. Jack Standish gave him help and protection. He couldn't leave them, even though the call of space beckoned to him like a lost lover.

But if he didn't leave now, and more Zom ships arrived, he could find himself stuck on Earth. He had to leave ahead of any Zom blockade.

Earth had made its own fate, the U.S. had sealed its own doom, and Stan knew that nothing he did could change the outcome. In three months, maybe less, the Earth he knew would be gone forever. If he stayed, then he might get himself and his ship sucked in and crushed by the closing vortex of events.

He had to leave. He should set course for Proxima Centauri right now and not look back. He wondered if he could use Earth for a jump-point, if that could give him an extra boost and get him there a little quicker. He turned to the navigation computer.

Then he saw a light through the view port.

It startled him at first, until he realized what it was, and then he whistled when he recognized it. He had never before laid eyes on the Northern Lights directly—he'd grown up in California, and lived in Florida much of his adult life. Why had he never visited the north?

The columns of light, the way they moved and shifted, transfixed him. He tried to look away, to resume his train of thought, but his eyes froze on nature's greatest light show as it played out in the cold night sky. He wondered what the nomadic people of old must have thought, seeing what must have looked to them like the shadows of the gods, dancing around mystical campfires.

He'd returned to Earth thinking that Saturn was the most beautiful thing imaginable. Now, watching the Northern Lights, he struggled to remember what Saturn even looked like.

All his life he had yearned for space and all its beauties and wonders. Yet here, right on Earth, was one of nature's greatest spectacles. The physics of magnetic field lines, charged particles, and excited molecules, though fascinating, paled under the shear glory of the result, generating thoughts of the supernatural in the mind of the scientific explorer, Stan McPherson.

"The world is still worth saving," he said to himself, "even if it's already too late."

His cell phone rang.

"Hello," he said.

"You don't know me," said the voice at the other end. The person had a thick British accent. "Mitch Ratcliff gave me your number. I have a package for you from a rather outlandish fellow of influence...a government official representing some mutual friends of ours."

This had to be a British intelligence agent—he'd referenced Stan's friend, Jack Standish, by his code name "Mitch Ratcliff". "Outlandish", in this context, could only mean alien. "Mutual friends" had to mean the Crenum, since they were "outlandish", and friends of the British.

So a Crenum government official was trying to relay a message to him through the British Secret Service...who'd been in contact with the Crenum for years through the local alien intelligence commander, Colonel Grandin.

"Ok," said Stan. "You should know that I don't hang out with that particular crowd anymore. Their activities turned out to be...unhealthy...for them, and for me."

"Ya, Mitch told me about the falling out you'd had—bad business. Why don't you drop on by Stegglehead tomorrow night...I'll meet you there and we can talk and I can give you this stuff."

"Stegglehead" was a code word that Jack had arranged with Stan. It referred to a specific coordinates off the coast of Norway. They had set it up as a possible pick-up point—should Jack somehow end up stuck back in England.

"Ok," said Stan. "I'll see you there."

"Good! Cheerio!"

Stan turned his attention back to the sight in front of him. It seemed like he was flying right into the lights, like they were a glorious new dawn of hope...

...almost like heaven itself.

He saw the steep, rocky shore of Coat's Island loom ahead and adjusted his course to skirt around to the east of it. He wondered if the Aurora Borealis would be out again tomorrow night. He looked forward to it.

He'd stay on Earth a little longer.

***

"Contact!"

The Crenum destroyer, *Stefler*, came to alert.

"Display it."

The large holographic viewer in the center of the bridge showed the red lines passing them at hyper-stellar speed. Then, more appeared as they came within range of the Stefler's graviton sensors.

"Illuminate them."

"Aye."

The plots multiplied as the powerful, Wide Graviton Beam transmitters aboard the *Stefler* came alive and began coursing through the void. The beams bounced off their targets and returning data on mass, course, speed, and other things, far in excess of the range of the *Stefler's* simple, quiet sensor package.

"They're headed for the Castlar cluster. Shall we alert them?"

"No, they'll have our scanner echoes by now."

***

"Admiral."

"I see them. Launch Task Force 20. Alert all other forces, already in flight, to converge here on the outpost and prepare to defend it. Order all civilian shipping to either leave the system immediately, or retreat here with all speed."

The admiral closed his eyes.

Into the Dark

Escape of the Nomad

So it begins—the blood, the death, the destruction, the broken minds and bodies. The war they'd all prayed would never happen had come.

"Alert Crenumar. Tell them we're under attack."

"Aye, Sir."

***

Admiral Tass watched as the huge, four ringed planet, Castlar 4, slowed to a stop alongside his flagship.

"Status."

"Twenty eight Crenum warships and a few civilian freighters are heading for Castlar 3. Ten other warships are already there. It looks like they got the warning from that destroyer back there and had time to abandon the gas planets."

"Compliment?"

"I see three carriers, fifteen cruisers, and twenty destroyers."

Odd.

"How many destroyers?"

"Twenty, Sir."

"That's not half enough pickets for all those heavy units. Scan the system carefully; they have to be somewhere. Pull our carriers and transports away from this planet until we're sure there's nothing hiding down there in the clouds, and re-scan those cruisers. Are any of them carrier killers?"

"We double checked them already; there are no attack cruisers among them. That's a good thing right?"

"Check them again, send in probes if you have to. There should be attack cruisers here, and not knowing where they are is *not* a good thing—not at all."

"Aye, Sir. Twenty or so ships left the system just as we arrived, five or six of them were large vessels."

The Crenum had presented him with two separate puzzles. Where were the attack cruisers? Where were the other fifty to sixty destroyers that should be here to cover those eighteen larger ships? Attack cruisers were the Crenum answer to Zom fighter superiority. They were swift, tough, and deadly—heavy cruisers with double mass drives and all their weapons mounted forward. "Carrier Killers", designed to pulverize destroyer pickets, crash through the defensive ring of cruisers and then cream his carriers. For the Crenum to neglect to post attack cruisers at a forward outpost like this one would be a stupid lack of foresight, but even that's not as bad as having such a thin layer of destroyer pickets. That would just be outright insane.

"Hmm," grunted Tass. "There should be carrier killers in this system, and their escort destroyers, and three times as many other destroyers and missile frigates besides. The Crenum aren't idiots—find me those ships! What were the vectors on the ones that jumped out?"

"We didn't map them yet, Sir."

"I know—they aren't here, so what do we care where they went? Map them!"

"Yessir."

He had a hunch. If he was right, then this whole attack may have been a huge error.

"Sir, we're getting reports back from the other two assault groups. There are no attack cruisers defending their targets either."

Damn!

"Sir, we have those departure vectors…"

"You are about to tell me that some of those heavier ships that left the system, and about four or five times that many lighter vessels, are currently enroute to Telesia?"

"Yes, Sir. Two heavy, ten light."

"Two attack cruisers, with five destroyer escorts each. Contact the *Racthra* in Telesia. Tell them to prepare for action against Crenum fast attack groups!"

All three Crenum outposts had sent their attack cruisers to Telesia, the moment they'd seen Zom forces enter Crenum space.

Captain…*Commodore*…Gibrack was going to have his hands full.

Chapter 17

"We can defeat a force that size, using flanking attacks," said Commodore Gibrack's tactical officer Greel Deriff, "but we can't maintain the blockade. I've run hundreds of simulations, the results are the same; we can't hold the outer system against six Crenum attack cruisers."

"We can't hold Telesia," said Commodore Gibrack, "but neither can they. Attack cruisers are practically useless for orbital defense."

Lt. Commander Greel Deriff had been with him through many engagements in the long and dangerous war against the Rodkley Collective. He valued his opinion highly. Deriff was a can-do type of officer, a very resourceful man with an almost miraculous talent for finding a way around difficult problems. If Greel said they couldn't do something, then it simply couldn't be done.

Fine then. Let the Crenum have the outer system. Our target is Telis 3.

"Our fighters have been raiding Telis 3 for seven days now—with zero losses," said the Commodore. "We have tactical air supremacy over the savages; nothing can fly unless we let it."

"Still, Sir," said Commander Miral, Gibrack's political liaison officer, "no grounded planet has ever killed so many of our people in a single day."

Commodore Gibrack snorted.

"Shellar's stupidity killed those people as much as those silly drones. If he'd let me send in fighters against the infernal things when they first appeared, we wouldn't be having this discussion. You know what I think, Miral? I think those drones were the Telesians' only trick, and now that we know about them, they're far less dangerous than six hot Crenum attack cruisers."

He hadn't heard back from Tass yet, so if he wanted to avoid or delay action with the Crenum, he either had to run to Telis 3 and begin the invasion, or run home.

He pulled up a map of the planet.

Miral started to say something else, but Gibrak interrupted him with a look. Commander Miral's presence at the meeting was a courtesy only—he was a lawyer, not a soldier. In Gibrack's experience with Miral, the man nearly always chose exactly the wrong options on anything military.

"What about their drive rings?" Miral put in with emphasis.

What about them? Thousands of Zom had died in this operation already—they couldn't just leave. Things had gone too far for that. The Telesians would certainly join the Crenum Alliance if the Zom withdrew now, and that would be a disaster. Even worse, the Telesians might come away from the experience thinking their stupid, desperate, low-tech drones had chased the Zom away, and that would make them insufferable neighbors forever. Besides, if the UCP ever got wind that Shellar had continued his approach after seeing the probe's drive rings on his scanners, it would show the Zom to have violated the Mistaac Accord.

Or would it, really? It's not as if those suicide drones could operate between star-systems, they used a light-speed triggering device for star's sake! Those drones *weren't* interstellar weapons, so their technology, though it might mean something to the diplomats, meant nothing to him.

And he intended to make sure no one else ever got wind of them.

"What drive rings?"

Miral closed his mouth with a snap.

"We'll proceed with the invasion," said Gibrack, holding his gaze on Miral a moment longer than necessary, to emphasize the point.

"We'll land our transport here," he said, "on the southernmost end of the continent of Greenland, right between our two main targets, the United States and Great Britain. Basing from there, our fighters can continue action against those two countries and still be close enough to help defend the orbiting fleet from the Crenum. Plus, it'll lure any hostile Telesian forces out of hiding so we can finish them off. The Crenum obviously sent those cruisers here to distract us, but we'll just take what we came here for anyway. If we can't hold this one planet until reinforcements arrive, then we deserve to lose Telesia."

***

"Admiral Tass, they just sent out a general alert."

"They must be recalling their fighter patrols. A general broadcast would do that without telling us where those patrols are."

"Well, it'd be nice if our larger ships could operate under stealth like fighters and not give them so much warning."

"Careful what you wish for, Commander," said Tass. "The Crenum usually run slightly ahead of us in technology. If ships larger than small freighters could use stealth, then the Crenum would more than likely develop them first…and the first *you* would likely hear about it would be a surprise attack on *us*."

"Of course, Sir."

Tass' own words gave him an idea. He still hadn't found the missing Crenum destroyers, and the *Cladnic* was said to be a stealth-capable craft—using Crenum technology—and it wasn't that much smaller than the smallest destroyer.

"Have we recalled the probes we left behind to watch the outer planets?"

"A few."

"Better bring them all in," he said. "Leave thirty or so, but service the rest as quickly as possible and prepare them for launch. We might need them later."

"Yes, Sir. What about watching the outer system?"

"If there's anything hiding out there, then it's hid from our close scrutiny for two days already. Everything in the system will join us here soon anyway, to defend the outpost."

***

The powers that be had given Lieutenant Frestil one of the most nerve-wracking jobs in the fleet—monitoring the stealth shields on her destroyer, the *Josort*. Destroyers weren't even supposed to have stealth abilities, and the technology was still temperamental. The worse part of her stress level came from the situation today—and the lethal consequences of failure.

The general alert from the outpost had been the signal to come out of hiding and begin their approach. If all went well, this war with the Zom would start with an extremely unpleasant surprise attack against their fighter carriers.

But the Zom admiral wasn't making it easy.

Into the Dark
Escape of the Nomad

The original plan had been to hang back and attack the carriers at a safe distance from the rest of the fleet, just as they reached striking range of the outpost. That would have forced the Zom to fight two separate battles at the same time.

Instead, the Zom had ignored their own military doctrine and brought their carriers right up close to their fleet, and then flanked their rear with extra cruisers. That would make it easier for the warships defending the outpost to hit the carriers, but it also made a sneak attack against the carriers' flank a lot more dangerous for the sneakers.

While she nursed her ship's hastily installed, overly experimental stealth equipment, her ship and fleet had to stalk their targets practically right under the guns of an invasion force powerful enough to neatly pulverize them.

She went through the checklist again, monitoring the temperature and flux readings for each component.

Odd.

Modules 9 and 10 still displayed the same temperature reading. The other components all showed slightly different temperatures, as they each heated up at different rates. Module 9 was one of the power transformers; Module 10 was one of the four Specto-Shift zones. The two components had very different thermal properties and she could think of absolutely no reason why they would mimic one and other's temperature.

She got out her diagnostic equipment. She should report this discrepancy, but what if the parallel was a coincidence? The officers on the bridge had enough to worry about; she didn't want to look like an idiot starting an alert that was probably pointless. She'd investigate further, and try to find something substantive to report. She carefully opened the service hatch on the shield generator and gazed inside.

One glance was all it took. Both the primary and the backup temperature sensor connections for the transformer were open—not connected to anything. She looked over at the Specto-Shift module and found two separate sets of data-link cables connected to its temperature sensor. Apparently, in the rush to install the unit, someone had inadvertently connected the transformer temperature monitor to the Specto-Shift module by mistake. That wasn't too big a deal, a simple oversight, except that she still had to watch the transformer's temperature.

She set her tool case on top of the shield generator, and got out her diagnostic meter. She quickly configured it to thermo-sensing, tested it, and connected her data-link leads to it. Then, holding the meter in one hand and the data-link leads in the other, and careful not to touch anything else, she reached inside and pressed her leads into the temperature sensor connectors on the transformer. The meter chimed to signal a successful connection and she glanced at it.

Good! The transformer's temperature was well within spec.

She breathed a sigh of relief and screwed the leads down tight, then set the meter in a safe place inside the generator, where she could see the display easily though the service hatch. An occasional glance at it would suffice to monitor the transformer until she could break someone loose to help her fix the misconnected sensor cables.

Time to call in, her commander would no-doubt consider this issue a top-priority item, especially under these circumstances, and procedure required more than one person for such a repair. She activated the communicator on her wrist, as she did, her hand brushed against her toolkit and it slid toward the open service hatch. She grabbed for it with both hands, pinning it to the edge of the hatch, but it flipped, spilling it's many metallic contents right into the gaping hole.

She winced as she listened to her tools bounce around inside the stealth shield generator.

Nothing happened.

She breathed a tentative sigh of relief and looked. Miraculously, everything appeared to have fallen in harmless places. She looked over at the power coupling on the transformer and gasped. A torque spanner teetered precariously on it, almost shorting out the power coupling against the unit's frame.

"*Lieutenant?*"

The sudden sound from her communicator startled her and she jumped, causing the empty tool case to fall inside the generator.

She watched in horror as it hit the transformer and bounced against the spanner—which in turn slid against the power coupling and exploded in a shower of sparks.

***

"Contact! Dead astern of the *Tifra*! A Crenum light-destroyer…Gish class!"

Admiral Tass looked down at his monitor. The warship had suddenly appeared on graviton sensors—completely out of nowhere. No ship could have gotten that close undetected—except under stealth.

"Launch probes behind the carriers, now! Broad dispersal!"

The picket destroyers, and all five heavy-cruiser escorts at the rear of the fleet, sent out their full load of several thousand probes, extra eyes designed to locate, track and target stealth fighters in combat, but the intermittent contacts they now showed Tass sent a chill up his spine. No fewer than eighty Crenum destroyers and missile frigates, along with something like two thousand fighters, shadowed his carriers and their escort cruisers!

He had found the missing destroyers, in exactly the worse place they could possibly be. He watched as the enemy ships suddenly went to full speed.

“Stop the fleet! Launch all fighters!”

He'd searched the system for two days and found only a few stealthed Crenum probes and the occasional fighter patrol, yet the Crenum had managed to hide an entire flotilla of light combatants and then slip them right up close behind him.

Even with stealth shields, there was no way they could have concealed so many ships from him for so long. You don't just *hide* eighty destroyers and frigates from a determined search like that—stealth or not. Stealth destroyers and frigates were a clear and dangerous new advantage, but they also seemed to have found a way to hide an entire fleet from probes!

Two tactic-altering inventions—and one of them still an unknown.

“Contact our fleets invading the other outposts. Tell them to halt their attacks immediately and cover their transports and carriers. Warn them to watch their flanks for *stealthed* Crenum destroyers and large groups of fighters operating at extreme range.”

***

Captain Fridaday, of the Crenum attack cruiser *Torrent*, watched the large blue planet, Telis 8, slow to a stop beside his ship.

“Sensor scan.”

Into the Dark
Escape of the Nomad

"Seven Zom destroyers, looks like they're moving to intercept us. The other gas giants in the system have destroyers guarding them as well. The rest of the Zom fleet is closing on Telis 3."

"Any of the other cruisers here yet?"

"Nope, just us."

It was, of course, impossible for six separate attack groups, from three separate systems, to arrive at the same destination at the same time.

"Order the *Sark*, and the *Writhic* to withdraw in-system with the insertion packages and begin their operations. We'll engage the Zom destroyers with *Valient* and *Ralcore*.

"Yes, Sir."

"Send the signal to our friends on Telis 3, I want *The McPherson* out of there before the Zom can box in the planet with fighters and probes."

"Yes, Sir."

***

"There it is," said Standish, "a signal from a Crenum Cruiser called '*Torrent*'."

"Yay!" shouted Claire.

"Gimlar's message said that even though they were coming from separate systems, they should all arrive within a day or two of each other," said Stan. "So we should have a full escort by the time we get to Jupiter. What's the weather like topside?"

"It's a dark and stormy night."

"Tell our passengers to find a place to sit down. Put some of them in charge and tell them to keep the others out of our way. If we find any

Zom I'll need both of you undistracted. Claire, did you get a chance to train any helpers?"

"Ya, these scientist friends of yours aren't much to look at, but they learn fast."

"Good, then let's go."

They rose from the ocean floor and climbed slowly toward the surface.

Stan turned to Claire and Jack. "I don't think that *The McPherson* is an appropriate name for the ship anymore. When it was still my ship, I was willing to put up with it, but now it belongs to all of us, as well as to our passengers."

"You got something better?"

"I don't know…we're not likely to return for some time. We'll all be homeless—maybe we can start there."

"You're not going to name the ship '*Hobo*' are you?" Claire said with a grimace.

"I was actually thinking of '*Vagabond*'."

"Hmm," said Jack, "that term doesn't hold the honor it used to; it's grown to be synonymous with '*vagrant*'."

"Ya," giggled Claire, "just like 'Brit' is synonymous with 'self imposed English language expert'!"

Standish gave her a dark look.

"Someone has to be," he grumbled.

Claire pulled up a thesaurus on her computer console.

"Ok, let's see, Vagabond…tramp, drifter, beggar, hobo, vagrant."

"Look under drifter," suggested Stan.

"Wanderer, transient, *rolling stone*? Hmm."

"No," said Jack, shaking his head with a frown, "already overused. Check under 'wanderer'."

"Itinerant, roamer, rambler, gypsy, nomad. *Nomad*!"

"Why not?" said Stan. "The Zom are always calling us primitive savages, and we do move about looking for resources and security."

"Then *Nomad* it is," announced Claire.

"Change the ID on the computer to match, and let everyone know that the moment this ship leaves the water they'll be aboard the *Nomad*!"

"Aye-Aye, Captain McPherson," mocked Claire with a salute.

Their probes returned and reported that visibility was very bad below the clouds and that hurricane Vince extended vertically to well over 30,000 feet below an empty sky.

Nomad, amid sheets of rain and flashes of lightning, climbed slowly out of the ocean and rose from Earth for the last time, gaining speed as it dashed upward into the sky.

Claire looked back.

"I wish we could have done more to help."

"I know," said Stan, "but you and Jack did quite a lot."

"There may be hope yet," said Jack. "Just before we left to rejoin Stan, one of your Yank scientists told me they may have found a way around alien shields."

Claire looked at him with raised eyebrows.

"How?" she asked.

"I don't know, something to do with bacteria—it starts with an 'E'."

Claire thought for a moment, then her eyes went wide.

"E. Coli?"

"Ya, that's it, how'd you know?"

"Back at mountain base, mice got into one of the barrels of nuclear fuel the Crenum use. When they tried to use that fuel, it shut down the reactor and they had to repair it. The system uses an alien micro-organism as part of the process—maybe the E. Coli bacteria in the mouse droppings mess with their bacteria somehow."

***

Tass looked over the results of his invasion—it could have been far worse.

Through all those years fighting the Rodkley Collective, they'd neglected the Crenum too much. They should have watched them closer and seen these technology advances. For a hundred years, since the invention of stealth shields, there had never been a way to create an envelope larger than that of a small transport. *The Crenum, along with everyone else, had ships of various purposes that fit barely inside that envelope.*

Large numbers of destroyers, and worse, missile frigates, that hide in a system and then sneak up on formations under cover of stealth would phenomenally change the face of warfare, and make things far more dangerous for attacker and defender alike. Circumstance had handed him the makings of a double disaster. First, the mess in Telesia, then the damage that these hidden stealth forces had done to his invasion of the Crenum outposts. His operations teetered on the brink of failure.

He'd lucked out on only one thing. One of the Crenum destroyers had lost it's stealth system in time to warn him. He'd dealt with those destroyers, and fighters, but not before they'd wiped out vast numbers of his fighters.

The missile frigates, however, had complicated things. Equipped with heavier, longer range ordinance, and with destroyers helping them aim their weapons from practically under his arm-pits, they'd handily wiped out his rear picket destroyers in the first salvo and then went after the carriers. The five heavy cruisers defending his flank had to physically interpose themselves to block the missiles and he'd lost every last one of them.

Few of the stealth destroyers and missile frigates had escaped the weapons of his heavier ships, and the two thousand Crenum fighters involved in the attack were mostly gone too. He'd lost a forth of his heavy cruisers, and most of his fighters, but saved his carriers and transports.

The attack had hurt him, but he'd managed the Crenum better than his fellow commanders at the other outposts. Crenum stealth destroyers, running picket for fighters and missile frigates, had succeeded in surprising the other two invasion fleets, thoroughly obliterating one of them and fighting the other to a stalemate. He still held enough strength to take and hold Castlar. The Crenum would have to defend their system with a very dangerous deficiency in picket destroyers. Destroyers are the whiskers of any fleet, without them the Crenum cruisers and planetary defenses would engage him at a severe disadvantage.

Yet he was worried. Had the fighters that attacked him launched from the carriers and outpost, or somewhere else? He still had not solved the mystery of how the stealth destroyers and frigates had hidden from his probes. He had to find out. Could they hide carriers in the outer system as well as stealth ships?

Perhaps Captain Thrigh, the Zom Governor in Telesia, could shed some light on this problem for him. The Crenum hid from him there for ten years.

Tass's fleet could probably take the Castlar outpost defenders—if their carriers and their fighter bases on the planet were now as empty as his carriers were, but if the fighters he'd destroyed had based from somewhere else in the system…

"Sir, the Commander of the outpost wishes to speak to you about a conditional surrender."

Tass smiled.

"Ours or theirs?"

"Theirs, Sir. They want to abandon the outpost and leave the system."

Things were looking up.

Chapter 18

"Contact! Strike 17 reporting a brief visual contact with an unidentified stealth craft, arcing into the light side of Telis 3 and around the back of the planet's moon." The fighter pilot replayed the sensor log to get a better look while he called it in.

"*Heading?*"

"Hard to say, its not on any direct course to anywhere, but it seems to be bound in the general direction of Telis 6 or Telis 7.

"*Try and I.D. it, 17, it might be some more of our people escaping the crackdown.*"

He found the image on the scanner log and zoomed it in. He almost couldn't believe his eyes, or his luck. He smiled.

"It's the *Cladnic*!"

"*After them! Reestablish contact! We'll have the fleet send out probes. Don't let them escape—remember what happened last time.*"

"I remember. Their luck just ran out."

He spoke with more confidence than he had. His fighter could catch the primitive Telesian ship easily enough, but the target was known to be well armed with something like a destroyer's worth of Crenum weaponry, and he'd only caught a glimpse of it as it headed outbound at it's top accel. He kicked his drive into full and steered his craft into an arced intercept course. Getting close enough to reacquire the *Cladnic* without the star, the planet, or the moon silhouetting his ship and giving him away might be tricky.

Yet, there were more than enough fighters in the area to take the *Cladnic* this time, if he could get good vectors on it. The problem would be covering enough space to box it in and locate it.

"Control, can you spare us some fighter pods for support? I think we'll have to work at extended range from the fleet during the search. We might be at this for a couple of days.

"*Agreed, Strike 17, pods and more fighters are in bound to you right now.*"

***

Commodore Gibrack listened to the combat reports. His destroyers had engaged the first Crenum attack cruiser and damaged it. He'd lost good people, throwing destroyers in front of a carrier killer like that. He'd wanted his first action against Crenum forces to go the other way around, but those cruisers could outpace his fleet, and he had to keep them away from him long enough to reach Telis 3 and land his transport. After that he would recall all of his destroyers.

***

"…and so we, the Governing Council of the Crenum Alliance, regretfully proclaim that a state of war now exists between the Crenum Alliance and the Zom Republic. This state of war shall continue, at least until the Zom Republic fully withdraws from our border outposts…and Telesia. We regret the circumstances that have led to this state of affairs. We are a people of peace, with war thrust upon us by others. Let the interstellar community judge us as they will, but the defense of our star-nation, now in great peril, is our foremost concern at this time."

A murmur began in the Assembly Hall.

"Quiet, please," said President Shakree. "Let Klicktah continue."

Klicktah put away the prepared statement and cleared his throat. His heart trembled within him. He wished things had worked out differently. He doubted his people had fully prepared for this war. Would he live to see its end, or would it squander the lives and resources of generations to come? He steeled himself to the task at hand. His people needed allies—desperately.

"We will take back Castlar," he said to the silent assembly, his aged voice crackled with defiance. "The Zom forces holding it, though well led, took heavy damage in the conflict and will be easily displaced. Telesia, however, is a different matter, one that this entire body is responsible for. We call upon all peoples who value the provisions of the Mistaac Accord and the UCP Charter, to think carefully about the events that led to this, and to ask themselves what they can do to help us reverse it. We ask for civilized star-nations everywhere...to combine with us...as we wrench the Telis System from Zom hands. They have more than proven that they cannot manage that system responsibly, and the UCP should watch over its continued development—after the Zom scourge has been forcibly removed from the area. That is all."

President Shackree stood as Klicktah sat down.

"We'll take a break now, so that the delegates can confer with those they represent concerning this tragic turn of events. I disagree with Klicktah however—we have proper legal proceedings that are underway right now to *peacefully* address Zom activities in Telesia. Please, people...please keep a clear head and don't expand this conflict. Many

Zom and Crenum lives have already been lost, both in Telesia and the Crenum outposts along the border. This session now stands in recess."

***

"What was that?" said Stan.

Jack replayed the visual scanner records.

"Alien probe…Claire?"

"Yep," she said with a sigh, "it's Zom."

"Do you think it saw us?"

"Saw us? We practically ran over it. If you'd been smiling it could have counted your teeth."

Stan changed course ninety degrees, and down, away from the ecliptic.

"That won't work this time," said Claire. "It'll have us locked in, along with who knows what all other probes and fighters in the area. You might be better off to just make a bee line for Uranus and see if we make it."

They had had to pass up Saturn, to evade detection by a fighter pod. That had meant an additional three days travel to Uranus. They just needed a few more hours to get there, then they could jump out of the system, and all of these cat-and-mouse games would be a memory.

"Jack," said Stan, "kill that probe."

Clair tried to say something, but Jack already had the probe zeroed in and a missile ready. It's transmissions ensured its demise.

"Now everyone knows where we are," said Claire.

"Exactly," said Jack, "the Crenum too."

"Ah!"

***

"The data from Telesia seems to point to evidence that the Crenum have been hiding on the surface of the gas giants themselves, but Governor Thrigh and his people could never confirm it."

"It's little wonder," said Admiral Tass. "As soon as they detected the search efforts, they would have just snuck away. Still, the possibility of surviving in such a place…"

"We should search the gas giants here, Sir."

"We'll have a hard time doing it effectively, with our shortage of fighters," said Tass, "and they probably scuttled anything that sensitive before they left, but we have to be sure—I don't want any more surprises. Surround each planet with probes during your search, in case you scare anything up."

"Yes, Sir."

***

"We can't let them stay on Greenland," said President Atequipa.

"We have a battle plan already in place, Sir," said Withim. "Communication has been tricky, without our satellites, but all available forces are headed toward the area. I suggest we coordinate our assault with that of the Crenum cruisers."

"Of course. Have all the Crenum ships arrived in the system?"

"Yes and the entire group left Jupiter early this morning. ETA here, thirty-nine hours."

"What forces do we have that can reach Greenland by then?"

"Well, we have the 1st in France. The Zom haven't found them yet, and they've just finished their refit. They'll use Vince to evade any

Zom fighters that pursue them after their strike. Besides them, we have two boomers, *Alabama* and *Kentucky*, and the *Theodore Roosevelt* Carrier Strike Group. The British Navy is sending out two carrier groups as well. It would be nice if those boomers could use their nukes though."

"No, I think it's better that they don't. I'm tempted, but we have the down-wind fallout to consider. Our new Crenum ambassador tells me that what they call 'living planets' are rare and precious, and aliens on both sides take a very dim view of anyone damaging an ecosystem with nuclear weapons."

Withim nodded. "We can't take our system back without help from the Crenum. We can't afford to offend them."

"Exactly, and we don't want to do anything that will reduce this system's value to aliens in general either, or make ourselves appear less enlightened in their eyes, it would undercut the Crenum's support from other star-nations. Do the Zom seem to have any other transports besides that one they landed in Greenland?"

"No," Withim shook his head, "not after what we did to them. Funny thing though, they didn't even bring a fighter escort when they came in. They just dropped straight down and landed like they already owned the place. They seem to think they can go back into the business of under-estimating us."

The President grinned.

"Good," he said.

***

"Here comes another one!"

The Zom fighter was making its first pass, firing a plasma missile from close range. The ship shook from the impact, but the shields held.

As the fighter fled, a plasma missile from *Nomad* chased it down and destroyed it.

"That one was a better shot; good thing he only lived to shoot once," Standish's face had turned red from mounting frustration and anger.

Claire's voice came over the intercom.

"I'm holding the shields up with spit and bailing wire down here, Stan, we've got to do something different, those fighters are killing us faster than we are killing them."

"We can't outrun them, killing them as they come is all we can do."

"Those two other fighter pods just entered strike range," said Standish.

"Great, that means ten more fighters."

"No, it looks like they've been bloodied up already. They've got some damage, and only five fighters between them."

"Still, add their fighters to the two here and that makes seven," Claire added. "Stan, we can't keep this up."

"Be thankful they can't spare anything much bigger."

The two remaining fighters attacked together, firing their missiles at short range on a high-speed pass. Countermeasures from *Nomad* caused both to miss, and the fighters darted, sluicing to evade *Nomad*'s return fire before curving around for another shot.

"Stan, that Zom patrol ship we picked up earlier is almost within weapons range, and two destroyers just showed up at Uranus. They'll be in range in three minutes."

"Alright, save the plasma missiles; we'll have to just use particle cannons on the fighters and patrol ship for now. I'm slowing the ship to delay contact with those destroyers."

“We can't slow down,” said Claire. “If we make it out of this alive, we'll need that momentum when we jump from Uranus or we'll lose days on our arrival at Sisna, we haven't enough food for the trip as it is.”

“We might be able to shoot our way past those destroyers to Uranus, but not if we have to fight the patrol boat and the fighters at the same time!”

“Wait, wait, Stan!” interrupted Standish, “The computer just ID'd those destroyers—they’re Crenum!”

***

The crew of the destroyer *Sark* saw the fight on their scanners; or rather, they saw the missiles, debris, and fighter pods.

“Those are Zom and Crenum missiles.”

“Look! Isn’t that the Telesian ship we’ve been looking for?”

“Ya, let’s get those fighters off them.”

“Tell *Writhic* to go in behind us.”

***

“Wow, those destroyers just toasted both the pods, and the patrol ship, in one salvo,” said Jack, leaning over his console. “The fighters are going after them, shouldn't we help?”

“No,” said Stan, looking back at the battle, “they did that to give us our shot out of the system. Besides, by the time we could turn around and get back there it'd be all over anyway. It's time to jump. Say good-bye to home.”

Five minutes later, *Nomad*, battered and bruised, took two fast, curving orbits around Uranus and vanished into the dark.

***

Captain Forester saw the approaching continent of Greenland on his heads-up display as he closed range. The AG enhanced, air-launched cruise missile his fighter carried would reach its target a little over thirty minutes after launch. He gave no orders to the other fighters in his attack squadron. They knew what to do, and there was no sense in giving the Zom any more notice than they had to have.

Doesn't matter, he thought, *they probably have us locked in anyway. They're just letting us sneak in closer so they can spring their own trap and catch us.*

A light on his panel came on and he glanced at it, then he pushed a button on his stick and felt his fighter lurch as the weight of 4000 lbs of missile fell from his aircraft and rocketed forward toward its target.

He swung his fighter around ninety degrees and threw on auxiliary power.

His squadron had an appointment to keep to the south—with a storm named "Vince".

"They just launched missiles, then turned and ran. The missiles appear enhanced with anti-gravity propulsion systems, they're closing at eight times the speed of sound."

Gibrack smiled.

"They can't hope for much accuracy with those, not with their targeting satellites out. They must be armed with fusion warheads."

"The database says they haven't used nuclear warheads on that weapon class in over fifty stellar cycles."

"Ya, well, we all know the database isn't completely accurate, don't we," he said with a wink.

Should he have his fighters shoot down the pathetic things, or should he demonstrate to the primitives just how worthless any of their weapons really were, nuclear or otherwise, against the base's shields.

Still, he couldn't allow the savages to ruin their environment with radiation clouds. He had orders to take the planet intact.

"Send out drones to destroy them. Order the fighter cover to ignore the missiles and go after the enemy aircraft that launched them."

"Yes, sir."

Commander Frakes, of the 539 Assault Squadron, Royal Marines, looked at his watch. The drilling was taking way too long.

"Come on!" he whispered. "Hurry it up!"

He knew his nagging wouldn't speed things along any, the delay was caused by the limited technology of the alien equipment they were using. The "Particle Drills" as the aliens called them, captured from the Zom and operated by his engineers, didn't appear to have been designed for the "strategic geography" shields the Zom used here. They were of the "spacecraft boarding action" type. He had to give his men some sense of urgency though—if nothing else, it gave him something to do while he waited.

Finally, the drill pushed through.

"Let's move it! We've got eighteen minutes to deliver this nasty stuff!"

He was grateful to the Zom. In keeping with their typical pompous nature, they weren't bothering to patrol their perimeter, and he had no time for a firefight now. His men picked up the heavy boxes, his "delivery", and hustled them through the opening. Anti-grav lifts were out of the question from here on in, as was any other technology that the Zom might detect with their energy scanners.

He signaled the engineers to take the valuable drills away and withdraw from the island, they wouldn't be needed anymore.

He hoped.

"It looks like they'll make it to that storm before we can catch them."

The commander of the Zom fighter squadron smiled.

"As if that will save them," he said. He switched his weapons to graviton tracking.

Captain Forester saw the towering storm approach and closed one eye, so that if a lightning flash temporarily blinded one eye, he could switch to the other. He activated the de-icers on the air-intakes of his fighter, and slowed to subsonic speed in preparation for the buffeting winds. He glanced behind him, half expecting to see the plasma missiles already closing. Tropical storm Vince, formerly Hurricane Vince, had given them a great opportunity by swinging further North than most hurricanes, but then in the past day or so it had slowed, making the timing of their hide-and-seek maneuver a little too thin.

The Zom fighter commander gave the signal to the other four fighters in his strike force. The eighteen Telesian fighters were in range and disappearing from his visual scanners as they entered the cloudbank, but their graviton signatures still blinked a bright red. He activated two plasma missiles and fired.

Then his mouth dropped open in shock as the graviton contacts suddenly all winked out together. His and his comrades' missiles began first to wobble, then to veer from side to side as they struggled to reacquire their targets.

"Where'd they go?" said one of his men over the com. "Did the storm destroy them?"

"Couldn't have," said another, "those craft could survive a storm. Besides they wouldn't all just go away together like that...would they?"

Captain Forester grinned as he watched the other fighters in his attack group vanish from his display.

Great things these, alien stealth shields, he thought. *I could get used to this.*

"O-Eight-thirty, mark!" said Captain Brown of the *U.S.S Alabama.* He turned and signaled to his weapons officer, who pressed a button on his console. The submarine began rotating through her missile tubes, launching her loads in 10 second intervals.

"Tube 22 clear," reported the weapons officer, three minutes later.

"Dive. All ahead full."

His role in the battle completed, his job now was to protect his ship from reprisal attack. Tubes 23 and 24 still held his *illegal* ICBMs, but he had no orders to use them. He did the math—twenty-two tubes, with ten Tomahawk-KAG cruise missiles per tube. His two-hundred and twenty Tomahawks would begin their attack in nineteen minutes. He knew another boomer was closer, emptying her tubes even now, as his birds flew over.

"Good morning, Mr. Zom," he whispered.

Commander Frakes watched as the last big box of powdered rodent feces went into the alien fuel storage containers, and wondered how one amassed such a collection.

They must have shook the rat cages of every research lab in Britain.

His men closed the hatch. They had already concealed the dispatched alien guards in a nearby closet.

He took the dust mask off his face and stuffed it in his pocket.

"All right gents," he said, "let's get out of here."

"More missiles, Sir."

Admiral Gibrack looked up from watching the Crenum attack cruisers bearing down on him and glanced over at his other screen. His eyebrows went up.

"Where'd those come from?"

"Right out of the ocean, Sir."

He watched as the targets multiplied into the hundreds. They were issuing from two invisible points offshore from the base.

"They must be using submersible missile platforms, where are our fighters?

"They're still searching that storm for the escaped Telesian aircraft."

It shouldn't be this hard. The bigger challenge was supposed to be the Crenum, not the Telesians, but if he somehow lost the base on the ground, a victory against the Crenum would be pointless.

"They'll shoot down whatever missiles they can," he said, shaking his head. "the others will just have to hit the shields. Recall our fighters. Have them hunt down those submersibles and get rid of them."

Commander Frakes stood at the spot where they'd drilled through the shield. He took off one of his gloves and unzipping a pocket of his coat, retrieved a Churchill Crown from it.

"Here goes nothing," he said. He looked at the face of the coin for a moment, then placing it between his fingers, he flipped it out in front of him. It bounced off something invisible in mid-air and fell to the ground at his feet.

He stooped down and picked it up. It felt warm. He stood and waited for about half a minute, then flipped it again.

Again, it landed at his feet.

Missile #A5-4E-76-89-5C saw the approaching coastline and selected it's target, a defensive battery at the southernmost end of the base. It issued orders to the missile directly behind it to follow it in with a Horizontal Attack Maneuver. Then it transmitted an election packet, with a value of 0, relinquishing command of the sortie. The other missiles each generated their own election packets, announcing their status with a number. #A5-4E-76-89-5C sorted the list in reverse order, checked the top, saw it didn't win the election, and heard #EE-65-BA-42-33 assume command. The entire process took about a hundred nanoseconds.

Commander Frakes picked up the coin for the eighth time.

He wondered how the Zom treated their prisoners. He wondered if they even bothered to take prisoners.

"Sir," said one of his men, "the Yank missiles are here."

He looked south and saw the contrails flying low over the ocean.

He looked down at the coin and pursed his lips, then flipped it again.

It arced, unimpeded, to the ground.

He stepped forward and reached with his hand where the shield had been and felt nothing.

"Let's go," he said, retrieving his coin before he ran east, toward the shore.

"The base shields just went down!"

"What?" Gibrack glared at his com officer.

"The base shields have failed! Something went wrong with the reactor powering them! Whatever the problem is, it's cascading through the shield system reactors and power units throughout the base. The portable ground units are fine, but all the enclosures and outer defenses are powering down."

The main fuel system!

"Evacuate the enclosures! Tell all portable units to raise their shields and open fire on the incoming missiles as soon as their clear! Order all personnel to find cover outside and brace for attack!

He looked back at his display just as the six Crenum attack cruisers opened fire on his pickets.

"Begin flanking those cruisers…now," he said.

"Bonzi!" transmitted U.S. Navy Tomahawk-KAG cruise missile #4E-76-89-5C.

Then it detonated.

Epilog

Captain Tohall carefully landed his freighter on the battered airfield, and watched as a large number of armed people in four-wheeled vehicles came out of concealment and surrounded his ship. These rebels had somehow been issued Crenum vid-coms and were well organized; he had called ahead and been told exactly where and when he would be expected.

One of them stepped out of his vehicle and waved one hand in the air, apparently a traditional Telesian greeting. He smiled and waved back as he considered his situation. Then he sighed and pressed the button to lower the ramp on his cargo hold. He didn't like arms running, but the pay was good, and he wouldn't get his money unless he completed the delivery. Moreover, now his ship was equipped with the latest Crenum stealth shield system, quite a bargain, if he lived long enough to finish earning it. He just had to avoid being boarded by the Zom, or hijacked by these desperate, primitive Telesians.

Captain Forester signaled his people to board the alien spacecraft and unload its cargo. It was the third shipment he'd received so that night, and they were scheduled to unload and conceal four more just like it before sunrise. He smiled as strange looking crates came down the ramp. They contained alien technology that had been prepared far in advance; personnel shields and small arms, alien anti-air and spacecraft weapons, and extra power units and fuel to power them. Earlier, he had received weapons and more stealth shielding systems to install on his armored equipment and aircraft. It amazed him to find easy to read instruction booklets in each crate, written in English, and formatted exactly the same as he was accustomed to seeing from equipment issued by Uncle Sam himself. The booklets bore the initials "CS". It made it seem like the equipment really hadn't traveled almost fifty light years to get there.

Whoever you are, "CS", you speak English like you were born to it.

***

Another ship, a heavily armed Zom mini-freighter, now re-flagged *Copernicus*, equipped with stealth shields and manned by a Crenum and U.S. military crew, landed by night in a clearing in the Peruvian mountains.

President Atequipa, his family, his staff and their families all stepped out of the trees and walked toward the alien vessel, many of them gazing up at the alien spacecraft with fear and amazement. The President, still holding his five-year-old daughter's hand, stepped onto the boarding ramp. The ship's commander, Captain Norton, walked down the ramp and shook his hand.

"Welcome aboard, Mr. President."

"Thank you for coming."

He looked around at the jungle.

A government in exile.

"Sir, our time window is short."

He breathed a long and tired sigh. He stepped back down onto the grass and stooped down to take a tiny American flag out of his daughter's hand. He looked at it for a moment, holding it up against the stars and spinning it between his fingers, before handing it back to her and standing up.

"The stars are so different here," he said. "It feels weird, looking into a sky I've never seen before, I guess I'd better get used to it hadn't I?"

"Yessir. We should be going, we're vulnerable here. The Crenum have a group of warships standing by to escort us to UCP headquarters."

The President picked up his briefcase and boarded the ship, the others followed.

"We'll be back," he whispered.

***

Redliv 3, Sisna Cooperative –

The long awaited Telesian experimental transport, the *Nomad*, arrived today at the Gladstin spaceport on Redliv 3, amid cheering throngs of spectators. The air traffic authority cleared a special corridor for the heavily armed craft that has been at the center of so much controversy.

Though an experimental freighter by design, the ship bristles with weaponry, and is said to be almost the equivalent of a Crenum light destroyer in armament, with several kills against Zom fighters already to its credit. This fact did not go un-noticed by members of the Telesian Interference Commission, a multi-race committee of experts assigned by the UCP General Advocate's office to investigate alleged violations of the Mistaac Accord with regard to Telesia.

"We have much work yet to do of course, but I watched the ship as it came in and I did recognize Crenum anti-piracy weapons mounted on it," said one Commission member, speaking on condition of anonymity.

For security purposes, and at the request of the General Advocate's Office of the UCP, the vessel was brought to ground at the Sisna Space Defense Port, here on Redliv 3, and impounded by authorities immediately upon landing. All of the refugees were escorted off the ship, fed, and brought to various medical and residential facilities on base. The ship's crew, including the ship's designer, Stan McPherson, were permitted to remain aboard, on condition that they cooperate fully with the investigation, and refrain from public comment until the investigation is complete.

"All of the crew and passengers arrived malnourished from their tight rations, but are otherwise in good health," said Glif Maylar, head of the Commission. "I can tell you that so far the vessel design itself doesn't appear to be of Crenum origin, although exactly to what degree the Crenum shared technology with McPherson is still under investigation. The vessel also appears to have been damaged to some degree by weapon fire."

Some think that violations of the Mistaac Accord may have played a part in the stunning casualties that were inflicted on the Zom by Telis 3 planetary defenses in the early stages of the war.

The Zom Republic regard the Telesian vessel, and everything on board it, as their property, since Telis 3 is a Zom planet. This claim was made quite forcefully in an official complaint issued by the Zom embassy today, voicing strong disapproval with Sisna's handling of the matter, and requesting that the ship be remanded into Zom custody immediately.

The reply from the Sisna government was swift. "The UCP ordered us to seize the vessel, because it is evidence in an investigation. No one has any claim on it until the UCP investigators release it," the Sisna Alien Affairs Office said in a prepared statement. "After that, the craft will be returned to the Telesians who brought it here."

Donors have stepped forward from all over UCP governed space, offering assistance for the Telesian refugees. Cultural scientists will converge on the Redliv system soon, and will probably employ many of the refugees to help them interpret the Telesian Cultural Database that McPherson brought here aboard *Nomad.* In the mean time, the Sisna government will administer a fund, providing generous stipends to the refugees to support them while they work their way into our society.

***

The Mit Kerick Group

News and Commentary

Special Edition – Interstellar War

Sponsored by: Zifnar Technologies – No matter what you do, we are there.

For copies of this transcript, contact the Mit Kerick Group, StarNet address: MitKerickGroup

Mit Kerick (Host): Issue One – War Declared.

The Zom yesterday announced without equivocation that a state of war now exists between the Zom Republic and the Crenum Alliance. This on the heals of embarrassing losses they have suffered at the hands of both the Crenum and the Telesians. Not only did Crenum involvement allow the escape of *The McPherson*, now renamed *Nomad*, but Crenum fast-attack units, running escort for swarms of small cargo ships ferrying arms and refugees, have been hitting Zom forces at Telesian jump-points on a regular basis. Some analysts say that Crenum fleet action—and technology—may have also contributed to the second Telesian victory, the missile attack that destroyed the Zom base on Telis 3.

Question: What does this declaration of war change, Mr. Terril?

Ferr Terril (Chief Opinion Editor for the *Zom Information Exchange*): It was more of a political move Mit; some of the Zom's key allies are restrained by their laws from getting involved directly in conflicts that do not have an official declaration of war, a specific objective, and a timetable.

Mit: But can the Zom even have a timetable under these circumstances?

Ferr: Well, simple timetables are never reliable in war anyway; everybody knows that. The timetables that these governments require can take many pages to describe—if this happens then that…if that happens etc.

Mit: Chez, do you agree with this assessment, that this is an attempt to gain support?

Chez Faltral (Executive Editor for the *Interstellar Journal*): I don't think that that is what Ferr meant, those worlds he is talking about already support the Zom side of the war, at least in concept. This declaration of war just resolves legal technicalities that stand in the way of direct assistance.

Ferr: Exactly!

Chez: But I do not think that that is the real reason for the declaration. You see, Zom behavior from the start of this mess has been abominable. I do think they are having trouble gaining support…they were caught with their hand in the pastry can and now they need someone else to blame for it. The beauty is in the details, Mit; there was absolutely no mention in this pronouncement of interference doctrine, which is what all of this is really about. The way that this declaration is worded, it basically tries to call the Crenum actions an invasion of Zom space, saying that the war is necessary to defend themselves. The ability to cry victim is the only good thing they can do with this string of defeats…

Ferr: Wait a minute, the Zom now hold Telesia!

Chez: If you say so Mit. Today they hold Telesia, tomorrow maybe not. Telesia has changed hands five times since the start of the war, but we can call some of their failed objectives "badly placed victories" instead of "defeats" if you like.

Mit: Sal, what are these "failed objectives" that she's talking about?

Sal Difnar (Chief War Correspondent of the *UCP Review*): Just what every news outlet has been saying Mit; you probably already have a list of them ready to roll today.

(*Laughter*)

Mit: I do, but I'll display them as you call it. (*Laughter*) Also, since you're the expert, I asked you in case I missed something.

Sal: Well, the first thing is, I agree with Chez. To call these engagements Zom "victories" requires that you widen your definition of the term to ridiculous proportions…

Ferr: (groans)

Sal: I know that you love the Zom Ferr, but when the most flaunted military force in the region looses control of its border to an opposing force this badly, it's not a victory. I'm sorry…it's not.

Mit: So the first failed objective is containing the Crenum?

Chez: No, it's…

Ferr: Well…

Sal: Well, chronologically no, that was the Telesian Surprise causing…

Ferr: That wasn't a failed objective. The landing was eventually achieved.

Chez: Ya, just before the Telesians bombed it to scrap!

Sal: The failed objective was not the landing; it was the one-day takeover. The objective was not just to land, but to finish the Telesians quickly. That didn't happen. The Telesian Surprise did too much damage, and then the Telesians fought too well and held on long enough to obtain better equipment and entrench themselves. That was made possible, in part, by the second failed objective, preventing Crenum intervention, and the third, to take and hold the Crenum border outposts. The Crenum can launch focused strikes into Telesia at will and the Telesians are now fully defiant, and equipped with Crenum weapons, shields, and other equipment. The Zom also have other problems marching up the road.

Mit: We'll get to those "other problems" in a minute. All right Dif, where do you think the war will go from here?

Dif Lunard (Editor in Chief of the *Financial Gazette*): All over Zom space.

Mit: Excuse me?

Dif: Think about it. This is no longer the mere border skirmish that the Crenum announced in their declaration of war—the wording in the Zom declaration contains no such language. The Crenum gun-runners…

Sal: They are not all Crenum.

Dif: Sorry, my mistake, many are privateers.

Sal: Right!

Dif: The Crenum have shared their improved stealth technology with contracted privateers.

Chez: Yes.

Dif: Anyway, these stealth freighters will now start smuggling arms to all of the other Zom food worlds. The Zom will have to chase them in every corner of their space. Biological products, which before all this were already in short supply in the Zom Republic, may now become a crippling shortage. Add to that the increased supply needs in a time of war and you have a terrible situation for the Zom. If they concentrate their forces to fight the Crenum, they can't protect their jump-points from escorted gun-runner convoys. Those Crenum fast-attack cruisers are turning out to be excellent for hit and fade attacks—much like destroyers, only with muscle and a bad attitude.

(*Laughter*)

Mit: Question - Does this Declaration of war help the Zom? Dif.

Dif: Nope, as I said it just complicates things.

Mit: Ferr?

Ferr: If their allies come through and they win this thing quickly, I'd say it helps.

Mit: Chez?

Chez: Neutral, but necessary.

Sal: I agree with Ferr, if their allies come through fairly quickly there will be no effect, they will be able to import their supplies from their friends, but even their allies cannot do it for free. Zom space is very large, and moving and protecting so many transports for such long voyages will cost them dearly.

Issue 2: Interference Doctrine

Both the Zom Republic and the Crenum Alliance stand accused by the other of numerous violations of the Mistaac Accord, a UCP law that forbids interference with grounded cultures. The Zom are accused of invading a primitive world protected by the Mistaac Accord. The Zom are accusing the Crenum of assisting McPherson in constructing his spacecraft and in providing the nations of Telesia with off-world warfare equipment and technology. Ferr, explain.

Ferr: Well first, the UCP is not going to act on the Zom charges against the Crenum for now. The General Advocate's office rejected them for lack of proof. They have since published their complaint to the media. The Crenum complaint basically says, in part, that the Zom invaded a UCP protected planet. The Zom claim that the planet had rejected space flight technology and was past its prime; that it would never have become a star-faring culture anyway, and therefore are exempt from protection.

Mit: How are they explaining the Telesian Surprise?

Ferr: They aren't. They're saying that there is no way to measure the damage that the Crenum have done on that world by their presence there over…

Dif: Oh please!

Sal: The Crenum…

Chez: Really! That takes the…

Ferr: Let me finish…

Dif: The Zom have been interfering…

Mit: Let him finish.

<Crosstalk>

Mit: Let him finish.

Dif: This is just Zom propaganda, to deflect…

Ferr: It is not propaganda. It is on the record that the Crenum have been on that planet…

Dif: Oh, come on!

Chez: Yes they have, but…

Sal: They have but…

Ferr:…illegally…unregulated…

Dif: Mit please, let me respond to that…

Mit: Please, let him finish, let him finish.

Ferr: They have been there doing who knows what for over ten Telesian years. No one can ever be sure how much damage they've done. Isn't it amazing how a grounded culture just *happened* to come up with drones equipped with drive rings—just in time to weaken a Zom fleet for Crenum attack cruisers to jump in and pounce on them?

Mitt: Are you saying that the Zom are blaming the developments that are evident in the McPherson spacecraft and in the weapons used in the Telesian surprise…

Dif: Ridiculous!

Mitt:…blaming those apparent Telesian advancements on the Crenum?

Ferr: Yes. How else could a permanently grounded culture like that have acquired such things? They also used an off-world particle drill and off-world stealth shields in that missile attack, and somehow coordinated the launch with an attack by Crenum fleet units. The Telesians could not have achieved those results in the drone attack, or the missile salvo, without illegal Crenum assistance.

Mit: Dif, go ahead.

Dif: The Zom were there over that same period, and some of the things that I'm hearing…

Ferr: Hearsay…

Dif: Some of the things that I'm hearing are not yet officially released to the media. They are direct and they are brazen, and they are obviously illegal. The Zom have been doing these things for decades…

Ferr: Hearsay

…and they have been going on for decades and the Zom have been doing them, and not just on Telesia. Telesia is where they got caught because it's close enough to the border for the Crenum to gather evidence and make a case out of it.

Mit: Thank you; you just introduced the next issue…

Sal: Just a second, there is another point.

Mit: What is it?

Sal: Well, just as Chez pointed out earlier, the beauty is in the details. The provisions of the Mistaac Accord that we're always hearing about are the rules which lay out what kind of culture qualifies for protection. Both the Crenum and the Zom say they will lean on those details pretty heavily to justify their activities in Telesia.

Issue 3: The UCP

Tomorrow the UCP General Advocate, Steck Lertis, will issue her opening statements to the UCP general assembly. Her case is said to contain a very interesting exposé of evidence against the Zom with regard to Telis 3. In the time she has been in office, she has championed very few cases before the UCP, preferring instead to resolve issues in the various committees.

She is rumored to have been looking forward to addressing this issue before the entire UCP, and sources tell us that she is livid…not only at the alleged actions of the Zom, but also with the inaction of the UCP General Assembly.

(UCP General Advocate Steck Lertis from vid recording)

"I cannot believe that they would pass a law like this and then ignore such obvious violations of it. What they have done is like telling one child not to beat up the other child, and then going into the house, closing the curtains, and stuffing things in their ears so they can't hear the victim screaming."

Mitt: Question, what can we expect from the General advocate tomorrow? Ferr.

Ferr: A lot of hot headed drivel. This woman is bad tempered even on a good day, and so I don't know how anyone can possibly say that she's angry. How does one know? They say that people yell loudest when their argument is the weakest, and we can expect to hear some very loud yelling from Ms. Lertis. I'm glad that she brought up the subject of hearing protection; I'll be bringing some.

(*Laughter*)

Mit: Sal?

Sal: We can expect a very long and interesting list. We can also expect a solid case. Something about this General Advocate, she does not take cases unless they are strong, and I disagree with Ferr; I know her pretty well and I certainly can tell the difference between her usual nasty disposition and the anger that she has been showing lately over this issue.

Mit: Chez?

Chez: I agree with Sal, it will be interesting, and not everything came out in the closed sessions. There will be new stuff brought out in the recorded testimonies of Telesian refugees from the *Nomad.*

Mit: Have there been any leaks?

Chez: No.

Sal: No.

Ferr: There is nothing to leak.

Dif: Oh, yes!

Mit: Yes?

Dif: Yes.

Issue Four: More problems for the Zom.

For the past three years, the Zom have been fighting a limited border dispute with an alien race who is not a member of the UCP. The Zom and their allies recently signed an armistice with that race, called the Rodkley. Now, however, some Rodkley are said to be reconsidering that agreement in light of the recent problems that the Zom are having with the Crenum and with Telesia. How will this effect the overall situation? Sal.

Sal: It complicates things even more for them. The Telesians and the Crenum are on the Zom frontier at one end, and the Rodkley are on their frontier on the other. They may find themselves in a two-border war that could erase both frontiers completely.

Dif: The economics of it are that the more time ships spend enroute across their space, the more biological supplies they use. This is very bad for the Zom right now.

Mit: Is there the chance of an alliance between the Crenum and the Rodkley?

Ferr: <Laughs>

Sal: Oh, no, not a chance.

Chez: No.

Dif: I don't see how that would even be possible.

Chez: They make the Zom look like angels. They are very brutal, and savage. Instead of conquering and subjugating other worlds, they obliterate and colonize; instead of enslaving conquered races…they eat them. The Crenum, and everyone else, has to keep them at arms length and away from their worlds.

Ferr: She's right. No one wants anything to do with the Rodkley, and the Rodkley regard everyone else as so much protein. Any emissaries that the Crenum send over there would never be seen again, except maybe on a plate.

Mit: Predictions, Ferr.

Ferr Terril (Chief Opinion Editor for the *Zom Information Exchange*): More light will be shed on Crenum crimes as information trickles out of Telesia, but the information will come from Telesian refugees, through the media, not the UCP legal offices.

Mit: Dif?

Dif Lunard (Editor in Chief of the *Financial Gazette*): Investments in biological packaging industries will continue to increase in value, prompting many new and very much needed facility upgrades.

Mit: Good!

Dif: I knew you'd like that Mit.

Mit: Chez?

Chez Faltral (Executive Editor for the *Interstellar Journal*): There is much more in the way of charges against the Zom forthcoming, and not just direct interference with the Telesians, in the form of secret agreements and media manipulation, but also…assassinations. Stay tuned.

Mit: Assassinations? Interesting. Sal?

Sal Difnar (Editor in Chief and Chief War Correspondent for the *UCP Review*):

The Zom will take this war deep inside Crenum space soon enough for you to feature it on our next show, Mit.

Mit: Really?

Sal: Yep.

Mit: I'll be sure and inform scheduling. I predict that Cicdos Fillal, current president of the Sasnar Commonwealth, will resign that post very soon, and flee the planet, amid yet to be revealed charges of corruption.

Next time—Meet the Telesians!

Mit: That's all. Good-bye.

***

"Where do I begin? There is so much here."

General Advocate Steck Lertis stood before the UCP General Assembly. Every seat was filled and the aisles were packed with reporters.

"I do not recall a time in history when a represented member of this organization has acted with such open contempt for its laws. Not only did the Zom make illegal contacts and illegal treaties with a grounded culture; not only did their activity in the system aid and incite a grounded military force to build star drive technology into weapons…and use them, but they also proceeded to invade Telis 3, even after it was obvious that the Telesians had developed technologies that gave them protected status under the Mistaac Accord. Mr. Klicktah and his people have secured documents, images, and sworn testimony from numerous individuals, some native to Telis 3. I am here today to begin the process of sharing that evidence with this assembly. The evidence shows how the Zom used secret agreements, direct manipulation, intimidation, and even murder, to tamper with the natural evolution of Telis 3 in a deliberate effort, to stagnate them and make them exempt from protection under the Mistaac Accord. Then, when they saw that they could not control the Telesians, a people hungry for space travel and scientific advancements of every kind, the Zom used military might to brutally assault them.

"Furthermore, while Telis 3 was supposed to be protected by statutes passed by this council, this council sat in these comfortable chairs and watched, and did nothing. The only thing more shameful than the Zom activities in this matter has been the cowardice of those here. The Crenum had to do the UCP's job in Telesia for many years, and their involvement has now started this war, a war which they are now fighting alone, and they should not be. The Crenum are now, single handedly, enforcing your ancestors' votes, because Zom bribery and threats have bound and gagged you in your seats. Our ancestors passed the Mistaac Accord to avoid the repeat of a catastrophe, yet neglected enforcement of this law has put us in danger of a tragic repeat of history, and at this time threatens the very stability of the UCP.

"During these proceedings, I intend to bring evidence against the Zom Republic, and against certain members of this council; I also intend to vindicate the brave actions of the Crenum agents, whose efforts gathered this evidence and brought these facts to light."

"My files are open to you Mr. President. Please schedule your hearings into this matter." She bowed and stepped down.

"Mr. Klicktah," President Shackree turned to face the Crenum Ambassador. "I once told you that we could take no further action until you were prepared to supply evidence of a spacecraft being constructed on that planet, and admit to Crenum trespassing into Zom space. Are you prepared to disclose these things?"

"That and much more Mr. President. General Advocate Lertis has all of my files and they contain the information that you requested. I defer all future arguments on this matter to her, and stand ready to act as advocate for the Crenum Alliance in defense of our actions."

"Good, then I shall look forward to reading those files for myself. Yes, General Advocate, we will begin the hearings that you request, and will inform your office of the schedule. Mr. Trelleg, I will be expecting your rebuttal to the General Advocate's opening statement in the next few days. However I do take umbrage to her accusations against this Council, which has acted appropriately and in perfect accordance with the law."

Lertis stood up and re-activated her microphone.

"What is it General Advocate? You have completed your opening statement."

"Yes, Sir, I have, but I would like to respond to the comment you just made."

"You are out of order, sit down please."

"I'm sorry, I can't."

"What?"

"Sir, you are the head of this body and you have just expressed a blanket bias against one of the main points in this case, prior to hearing the evidence."

"I did no such thing!"

"Respectfully, Sir, you did. You disagreed with my allegation that this body has acted inappropriately. Then you publicly declared that the council has broken no laws. Since you have chosen to make these declarations here and now, I would like to disclose the evidence that I have with me on those particular subjects, and open debate on them…here and now."

"You are currently the only person here in possession of that evidence, General Advocate, no one else has prepared arguments on the topic."

"I'm only walking through the door you opened, Mr. President."

"It can wait for the schedule, Ms. Advocate."

"No, Sir, it can't. By then your comments will have already been distributed to the people, and engraved in the public record. You will be politically bound to them, Sir, and it will put my case on an un-equal footing."

"Your comments are now very much out of order Ms. Lertis; I am only trying to maintain the influence of this Council to maintain authority over such matters."

"So am I."

"Madam Advocate…"

"Whether or not this body acted in complete accordance with the law is still subject to debate. Whether or not individual representatives here have behaved in perfect accordance with the law is still subject to debate."

A cold grin spread across her face. "A debate which I am more than prepared to start…right now!"

She took out a data module and held it in the air as she shouted, "I accuse this body of cowardice, and I have evidence of it here in my hand! I accuse this body of shameful—and illegal—neglect of their own laws, and I have the evidence right here! I accuse some members here of bribery, of accepting bribes, of conspiracy, and of a variety of other sordid dealings, and I have the evidence right here! I have names, times and places. The UCP legal committee has seen this evidence and they support it! I have evidence of the existence of the McPherson ship, the *Nomad*, that was available to this council well before the invasion—and ignored. I have images, eye-witnesses, and sworn testimony of direct government-to-government contact between the Zom and the *protected* culture on Telis 3! I have some very old documents...detailing...secret deals not only between the Zom and Telesian governments, but also between the Zom and various other member nations of this council with regard to Telis 3! I have cultural science data that was collected directly from the crew of the *Nomad*, and made available to some members of this council before the invasion. I have images and sworn testimony of an armed assault by the Zom against *Nomad*, while it was still an experimental aircraft, flying over the oceans of Telis 3.

"Now, Sir, the populations of much of the UCP will view this session. Are you going to retract your rash statement, or do I insert this into my display console and air this evidence, in its entirety, to the public, right here, right now, with the consequences to this council and certain members of it be damned?"

"Madam Advocate..."

"Sir?"

Shackree paused. He looked across the assembly hall at the massive plaque on the wall, containing the UCP charter:

We gather here:

In the interests of freedom and peace.

In the interests of worlds, cultures, languages, peoples, and religions.

To secure these things for ourselves,

By protecting them for others.

He looked down at his desk for a moment, and then back at the General Advocate.

"I retract my earlier comments, Ms. Lertis, and I look forward to viewing your evidence," he said quietly. "Now sit down please."

She sat back down.

He pounded his gavel, "Now onto other matters."

***

"So, I see you're spending your allowance frugally," said Claire in a reproachful tone.

Stan was on the roof of the space dock, trying out his new telescope.

"I had to leave my other one behind at my house in California."

"Well, maybe I should go spend mine on toys too."

Stan looked up at her and winked.

"My toys can be stowed in the cargo hold more safely than yours can."

"When I was little, my dad taught me how to navigate using the stars," she said wistfully, looking around at the sky. "Not just the North Star, but the others as well; but this sky is strange."

"Not all that strange," Stan said. He stepped around behind her and put a hand on her shoulder, then pointed past her ear at a bright star on the horizon. She looked down at his hand and smiled.

"That one there," he said, "is Deneb."

"Really?" she said. "It looks the same."

"It is the same; I doubt we will ever travel far enough for it to look much different than that. It's a white super giant, very rare, very distant. The greenies tell me that they used it to navigate anciently, same as we. They still calibrate their navigation systems by it."

It was a chilly night, but the space between them was getting warm. He stepped away, red faced, and went back to the telescope. She turned and watched him, wrinkling her nose in annoyance.

"Where's Sol?"

He pointed without looking up, "About there, in the center of that…umm…kind of square thing."

She looked at it a long time.

"I wish we could go back," she said.

"We can't do anything for Earth by returning there now," said Stan, adjusting his telescope. "We need to move on. We'll have to fight for Earth by making ourselves one with the aliens; finding friends and allies."

She turned and looked where the telescope pointed. She leaned close.

"Whatcha lookin' at?"

"Our next destination, one of the newest members of the Crenum Alliance. They need armed cargo ships."

"May I see?"

Stan stepped away.

"Sure, it's the one in the middle."

She gazed through the telescope at the small, bright star.
"I wonder what their sunsets look like," she said.

Into the Dark

Escape of the Nomad

Twinkle, twinkle, little star,
How I wonder what you are!
Up above the world so high,
Like a diamond in the sky!

Twinkle, twinkle, little star,
How I wonder what you are!

Into the Dark

Escape of the Nomad

There are a lot of people I'd like to thank who had a hand in the making of this book.

My family, who let me neglect the house, the yard, the cars, and all the various other things that other men did for their families while I sat, sequestered away, writing.

My test readers deserve much thanks too, such are the secret backers that no writer can be without.

I would also like to thank Cowboy Logic Press for buying into my dream and putting ink to it, my illustrator Chelsea Conlin for the amazing cover, and my editor Renee Herbert who honed that extra edge of quality into the punctuation and grammar.

Also my publicist Doug Johnston for helping get the word out, without him you might not have had a chance to read this book.

I must thank you, the reader, too—for picking the book up and not setting it down. If you enjoyed reading it, then stay tuned. I enjoyed writing it and will soon publish more.

Lastly, I thank my God, for giving me such a beautiful and glorious universe to look upon and write about.

Clear skies.

Into the Dark

Escape of the Nomad

About the Author:

Retired computer instructor, astronomy enthusiast, and author of the cultural sci-fi short story "Another Man's Terrorist" in the 2008 anthology "Satirica", Bill Housley's stories combine fast action, political intrigue, and a host of cool gadgets into spellbinding science fiction adventures for the whole family.

"The human race grew through volcanic eruptions, meteoroid impacts, and an ice age. *Into the Dark* shows us at our best, scratching and clawing our way from the bottom of the food chain to survival and supremacy--the only way to explore fully the greatness of mankind."

For more of his work, please visit his blog at:
http://bhousley.wordpress.com

LaVergne, TN USA
21 September 2010
197706LV00004B/3/P

9 780981 685311